ENVIRONMENTAL CHEMISTRY IN THE LAB

Periodic Table of the Elements

Atomic Number → 1 H Hydrogen 1.008 ← Symbol; Name → Hydrogen; ← Atomic Weight

1 IA	2 IIA	3 IIIB	4 IVB	5 VB	6 VIB	7 VIIB	8 VIIIB	9 VIIIB	10 VIIIB	11 IB	12 IIB	13 IIIA	14 IVA	15 VA	16 VIA	17 VIIA	18 VIIIA
1 H Hydrogen 1.008																	2 He Helium 4.002602
3 Li Lithium 6.94	4 Be Beryllium 9.0121831											5 B Boron 10.81	6 C Carbon 12.011	7 N Nitrogen 14.007	8 O Oxygen 15.999	9 F Fluorine 18.998403163	10 Ne Neon 20.1797
11 Na Sodium 22.98976928	12 Mg Magnesium 24.305											13 Al Aluminium 26.9815385	14 Si Silicon 28.085	15 P Phosphorus 30.973761998	16 S Sulfur 32.06	17 Cl Chlorine 35.45	18 Ar Argon 39.948
19 K Potassium 39.0983	20 Ca Calcium 40.078	21 Sc Scandium 44.955908	22 Ti Titanium 47.867	23 V Vanadium 50.9415	24 Cr Chromium 51.9961	25 Mn Manganese 54.938044	26 Fe Iron 55.845	27 Co Cobalt 58.933194	28 Ni Nickel 58.6934	29 Cu Copper 63.546	30 Zn Zinc 65.38	31 Ga Gallium 69.723	32 Ge Germanium 72.630	33 As Arsenic 74.921595	34 Se Selenium 78.971	35 Br Bromine 79.904	36 Kr Krypton 83.798
37 Rb Rubidium 85.4678	38 Sr Strontium 87.62	39 Y Yttrium 88.90584	40 Zr Zirconium 91.224	41 Nb Niobium 92.90637	42 Mo Molybdenum 95.95	43 Tc Technetium (98)	44 Ru Ruthenium 101.07	45 Rh Rhodium 102.90550	46 Pd Palladium 106.42	47 Ag Silver 107.8682	48 Cd Cadmium 112.414	49 In Indium 114.818	50 Sn Tin 118.710	51 Sb Antimony 121.760	52 Te Tellurium 127.60	53 I Iodine 126.90447	54 Xe Xenon 131.293
55 Cs Caesium 132.90545196	56 Ba Barium 137.327	57 - 71 Lanthanoids	72 Hf Hafnium 178.49	73 Ta Tantalum 180.94788	74 W Tungsten 183.84	75 Re Rhenium 186.207	76 Os Osmium 190.23	77 Ir Iridium 192.217	78 Pt Platinum 195.084	79 Au Gold 196.966569	80 Hg Mercury 200.592	81 Tl Thallium 204.38	82 Pb Lead 207.2	83 Bi Bismuth 208.98040	84 Po Polonium (209)	85 At Astatine (210)	86 Rn Radon (222)
87 Fr Francium (223)	88 Ra Radium (226)	89 - 103 Actinoids	104 Rf Rutherfordium (267)	105 Db Dubnium (268)	106 Sg Seaborgium (269)	107 Bh Bohrium (270)	108 Hs Hassium (269)	109 Mt Meitnerium (278)	110 Ds Darmstadtium (281)	111 Rg Roentgenium (282)	112 Cn Copernicium (285)	113 Nh Nihonium (286)	114 Fl Flerovium (289)	115 Mc Moscovium (289)	116 Lv Livermorium (293)	117 Ts Tennessine (294)	118 Og Oganesson (294)

57 La Lanthanum 138.90547	58 Ce Cerium 140.116	59 Pr Praseodymium 140.90766	60 Nd Neodymium 144.242	61 Pm Promethium (145)	62 Sm Samarium 150.36	63 Eu Europium 151.964	64 Gd Gadolinium 157.25	65 Tb Terbium 158.92535	66 Dy Dysprosium 162.500	67 Ho Holmium 164.93033	68 Er Erbium 167.259	69 Tm Thulium 168.93422	70 Yb Ytterbium 173.045	71 Lu Lutetium 174.9668
89 Ac Actinium (227)	90 Th Thorium 232.0377	91 Pa Protactinium 231.03588	92 U Uranium 238.02891	93 Np Neptunium (237)	94 Pu Plutonium (244)	95 Am Americium (243)	96 Cm Curium (247)	97 Bk Berkelium (247)	98 Cf Californium (251)	99 Es Einsteinium (252)	100 Fm Fermium (257)	101 Md Mendelevium (258)	102 No Nobelium (259)	103 Lr Lawrencium (266)

ENVIRONMENTAL CHEMISTRY IN THE LAB

Ruth Ann Murphy

CRC Press is an imprint of the
Taylor & Francis Group, an **informa** business

Cover information: The agave succulent, a possible source of biofuels for the future, is paired with traditional laboratory glassware used in research for new, environmentally friendly energy sources.

First edition published 2022
by CRC Press
6000 Broken Sound Parkway NW, Suite 300, Boca Raton, FL 33487-2742

and by CRC Press
4 Park Square, Milton Park, Abingdon, Oxon, OX14 4RN

CRC Press is an imprint of Taylor & Francis Group, LLC

ISBN: 978-0-367-43937-8 (hbk)
ISBN: 978-0-367-43895-1 (pbk)
ISBN: 978-1-003-00656-5 (ebk)

DOI: 10.1201/9781003006565

Typeset in Times
by MPS Limited, Dehradun

Access the Support Material: http://www.routledge.com/9780367439378

Dedication

In Honor
of
My Students

Contents

Preface....xi
Acknowledgments....xv
Author....xvii

Introduction to Laboratory Glassware and Supplies....1

Unit A
Introductory Exercises

Exercise 1 (1.25 hours)
Safety in the Environmental Chemistry Lab....13

Exercise 2 (0.75 hour)
Ethics in the Environmental Chemistry Lab....25

Exercise 3 (0.75 hour)
The Laboratory Notebook in the Environmental Chemistry Lab....29

Exercise 4 (1.5 hours)
Introduction to Glassware in the Environmental Chemistry Lab....35

Exercise 5 (1.75 hours)
Exploring the Metric System....51

Exercise 6 (1.75 hours)
Mastering the Metric System....57

Exercise 7 (0.75 hour)
Parts Per Million (ppm) and Other Tiny Quantities....61

Exercise 8 (0.75 hour)
Big and Small Numbers – Scientific Notation....65

Exercise 9 (1.75 hours)
Temperature and Heat – What's the Difference?....71

Exercise 10 (0.75 hour)
Some Elements of Matcha Mole's Adventure in
Environmental Chemistry Land 77

Exercise 11 (1.5 hours)
Environmental Chemical Reactions – Balancing Equations 81

Unit B
Adventures in the Environmental Chemistry Lab

Exercise 12 (1.5 hours)
Measurement – A Foundation of Environmental Chemistry 89

Exercise 13 (1.5 hours)
Why Do Icebergs Float? A Study of Density 95

Exercise 14 (2 hours)
Metals and Alloys – What's the Difference? 103

Exercise 15 (2 hours)
Recycling Copper – Learning Reactions 113

Exercise 16 (2 hours)
Studying Evaporation and Preparing "Canned Heat" 121

Exercise 17 (2 hours)
Can We Get Energy from Acid-Base Reactions?
Exploring Thermochemistry 129

Exercise 18 (1.25 hours)
Are Electrolytes Changing Our Waters? Identifying Electrolytes
with Conductivity Measurements 135

Exercise 19 (0.75 hour)
Radiation, An Invisible Pollutant – Measurements and Calculations 143

Exercise 20 (1 hour)
Is Extremely Low Frequency Radiation Harmful? A Laboratory Analysis........149

Exercise 21 (1.5 hours)
Adventures in Organic Chemistry Land........155

Exercise 22 (2 hours)
What's in Red Cabbage? Adventures in pH Land........163

Exercise 23 (1.75 hours)
The Process of Extraction – Measuring Caffeine in Tea Bags........171

Exercise 24 (2 hours)
Soap or Detergent: Which Is Better? Preparing Soap........179

Exercise 25 (2 hours)
What's in My Detergent? Measuring Phosphate with Spectroscopy........189

Exercise 26 (1.75 hours)
What's in Water Besides H_2O? A Laboratory Analysis........195

Exercise 27 (1.75 hours)
Purifying Cloudy Water – A Treatment Process........203

Exercise 28 (2 hours)
Is This Air Okay to Breathe? A Laboratory Analysis........213

Exercise 29 (1.75 hours)
What Are Plastics, Anyway? A Study of Polymers........223

Exercise 30 (2 hours)
Should I Bother with Recycling? Separation of Plastics by Density........235

Unit C
Closing Statements for the Environmental Chemistry Lab

Exercise 31 (1.75 hours)
What Is Illegal about Pollution? An Environmental Law Presentation 249

Exercise 32 (1.5 hours)
A Soldier's Perspective on Environmental Stewardship 263

Biographical Notes 287
Glossary 289
Index 295

Preface

TO THE STUDENT

Congratulations on embarking on a study of Environmental Chemistry! Participating in a lab is a great privilege as you get to do experiments rather than just hear about them. This book was designed for *you*, the student. The exercises are designed to be interesting, fun, and to provide you with useful information – for use now as well as in the future. We begin with some basic scientific exercises, to build a foundation before moving on to more obvious environmental topics, such as air and water quality, etc. Most or all of you are probably not science majors; yet you have the opportunity to vote on environmental matters. May you use the information you learn from this text to be a better citizen, and to guide your voting for your own benefit as well as that of your family, your country, and our beautiful planet! By participating in the labs, you have the opportunity to learn how science is done. We are all in this together, as the saying goes.

It is very important to study the assigned experiment before you come to lab. Your instructor may ask you to write out a plan for the experiment in your lab notebook in advance. Try to answer the Pre-Lab Questions before coming to lab as well. They are designed to help you work knowledgeably and gain more from your lab experience. A glossary of some of the important terms is included for your quick reference.

In labs, results, called "data," are collected and recorded in a lab notebook. Units such as grams, milliliters, etc. should be carefully included. Safety principles must be observed and experimental proof is required before new ideas are accepted. And some of you just might decide to add a major in Environmental Chemistry, or Chemistry! Remember, Chemistry is fun, and, if anyone tells you it is too hard, just tell them, "Chem-is-try!"

TO THE INSTRUCTOR

These experiments have been tested over the years in many labs of mine as well as of other instructors, and can be done in almost any order. It is recommended that the sections on safety, the laboratory notebook, and glassware be used early in the semester. Part of the fun of being an environmental chemistry teacher is the opportunity to be creative. For that purpose, several "Worksheet Labs," which function well remotely as well as in the classroom, are included here. They can supplement and/or replace some hands-on labs, fill in extra time as enrichment or extra credit projects, etc. Two "Lab Lectures," which I have used similarly, are included. While the efficacy of hands-on labs is well known, sometimes circumstances such as building maintenance, insufficient supplies, and pandemic restrictions favor the occasional use of such replacement assignments. This text includes more than sufficient work for a one-semester laboratory course, allowing the instructor the option of pairing two activities for some labs. Estimated completion times for the labs (although these can vary with instructor preferences and circumstances) are included on the Contents pages. Supply lists for 24 students or teams are included with each experiment. The preparation of slight excesses of amounts required is recommended due to possible breakage, spills, etc.

While this text should work well with any one-semester non-science majors' Environmental Chemistry/Environmental Science lecture text, the following pairings are suggested for those also adopting *Environmental Chemistry in Society, Third Edition*, by James M. Beard and Ruth Ann Murphy, Taylor & Francis Group, Boca Raton, FL (2021). Multiple labs are listed for some chapters to provide the instructors with options.

***Environmental Chemistry in the Lab*, Lab Text**	***Environmental Chemistry in Society, Third Edition*, Lecture Text**
Exercises 1–4	Chapter 1 Background to the Environmental Problem
Exercises 5–8, 12–13	Chapter 2 The Natural Laws
Exercises 10–11, 14	Chapter 3 Underlying Principles of Chemistry
Exercises 21–22	Chapter 4 Types of Chemical Compounds and Their Reactions
Exercise 15	Chapter 5 Element Cycles
Exercise 23	Chapter 6 Toxicology
Exercise 17	Chapter 7 Traditional Energy Sources and Modern Society
Exercise 18	Chapter 8 Emerging Energy Sources and Modern Society
Exercise 16	Chapter 9 Weather and Climate
Exercise 28	Chapter 10 Air Quality
Exercises 19–20	Chapter 11 Indoor Air Quality
Exercise 9	Chapter 12 Global Atmospheric Change
Exercise 26	Chapter 13 Water
Exercises 24–25, 27	Chapter 14 Water Quality
Exercise 29	Chapter 15 Solid Wastes
Exercises 30–32	Chapter 16 Hazardous Wastes

A global emphasis with brief biographical notes describing some outstanding scientists has been included where possible to provide students with the perspective that science is an international endeavor which is often enhanced by collaboration.

To summarize, this lab text is designed with the following features:

1. Four options are given for labs.
   a. Traditional, hands-on wet labs.
   b. Traditional labs with sample results provided by lab assistant "Matcha Mole" so students can write their lab reports based on data provided. These results are available to the instructors who adopt this text.

   c. Worksheets that can be done in a classroom or remotely.
   d. Lectures with material provided by experts, to be followed with a quiz.
2. Traditional labs have a pre-lab data sheet to be completed, and post-lab questions; lab preparation guides, keys, and suggested grading rubrics are available to instructors who adopt this text.

Any suggestions or corrections for possible future editions of this text would be most welcome and could be sent to rmurphy@umhb.edu.

Acknowledgments

Many illustrious colleagues have enriched and influenced my teaching career. I am particularly indebted to Darrell Watson, PhD, and the late Morris Stubbs, PhD, who provided the inspiration for much of this work. The contributions of the "Lab Lectures" by Colonel Rick Hoefert, MS, MSS, and Mr. John Mayfield, MS, are also gratefully acknowledged. Appreciation is due to Mr. Larry Carothers, MS, for sharing his lab preparation instructions. I am also grateful to my granddaughter, Gabrielle Murphy Goldstein, for providing the back cover inset and the Matcha Mole drawing. The safety cartoon provided by my student, Morgan Raines, is also gratefully acknowledged. Kudos to Hilary Lafoe at Taylor & Francis for her assistance and encouragement.

Ruth Ann Murphy

Author

Ruth Ann Murphy, PhD, earned a BS in Chemistry and a PhD in Physical Chemistry from the University of Texas at Austin with additional graduate studies at the University of Wisconsin–Madison. After teaching appointments in Chemistry at Southwestern University, Georgetown, Texas, and the University of New Mexico, Albuquerque, she served as Chemistry Coordinator and Assistant Professor at the University of Albuquerque. As Professor of Chemistry, she then chaired the Division of Science and Math at Howard Payne University, Brownwood, Texas. Since 1995, Dr. Murphy has taught at the University of Mary Hardin-Baylor in Belton, Texas, where she chairs the Department of Chemistry, Environmental Science, and Geology, holds the Amy LeVesconte Professorship of Chemistry, directs the Recycling Program, co-chairs the Health Professions Advisory Committee, and is Principal Investigator for the Robert A. Welch Foundation research grant. She has served as President of the Texas Association of Advisors for the Health Professions and is a member of the American Chemical Society including the Divisions of Chemical Education, Chemical Health and Safety, and Fluorine Chemistry. She is the co-author with Dr. James M. Beard of *Environmental Chemistry in Society, Third Edition*, CRC Press, Taylor & Francis Group, Boca Raton, FL (2021). Dr. Murphy enjoys making chemistry relevant to her students by including environmental chemistry and other examples from daily life in her courses.

SAFETY

The experiments in this text require knowledgeable handling of the chemicals and equipment, and must not be undertaken by anyone without an appropriate foundation in and commitment to the principles of laboratory safety and appropriate disposal of chemical waste. Although these experiments have been done many times successfully and safely, proper caution is always required.

Excellent safety information is available from many sources including the Division of Chemical Health and Safety of the American Chemical Society, the Laboratory Safety Institute of Natick, Massachusetts, and the *CRC Handbook of Laboratory Safety, Fifth Edition*, by A. Keith Furr, CRC Press, Boca Raton, FL (2000).

Introduction to Laboratory Glassware and Supplies

Balance, centigram. Weighs to nearest 1/100 of a gram. The term "balance" is preferred to "scales."

Beaker. Note the lip for directing the flow of liquids.

DOI: 10.1201/9781003006565-1

Beaker tongs. Useful for protecting from burns when lifting beakers and evaporating dishes which have been heated.

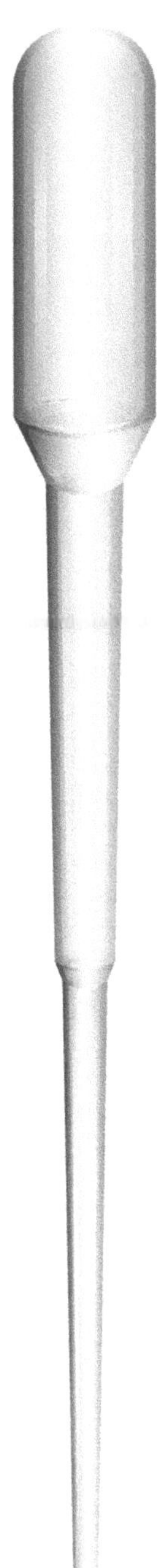

Beral pipet. Disposable means of transferring small amounts of liquids.

Buchner funnel. Suction from sidearm of filtering flask provides rapid filtration (separation of solid from liquid).

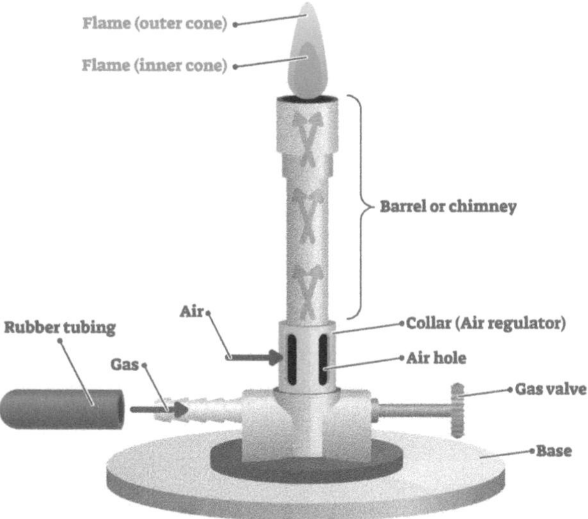

Bunsen burner. Air and gas adjustments allow for blue flame and efficient heating.

Centrifuge. Accelerates settling of precipitates when samples are placed in tubes and spun around.

evaporating dish

Evaporating dish. Convenient for removing excess liquid from samples when heated.

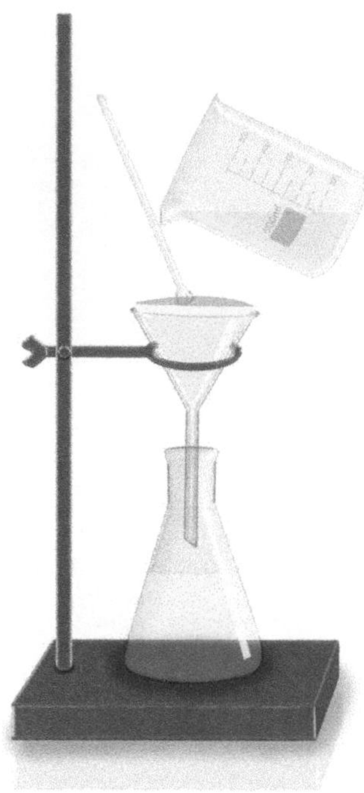

Conical funnel supported by ring stand for gravity filtration. Note use of glass rod to direct liquid to funnel.

Graduated cylinder. Useful for measuring liquids. Volumes in this 100-milliliter (mL) cylinder can be estimated to the nearest tenth of a milliliter.

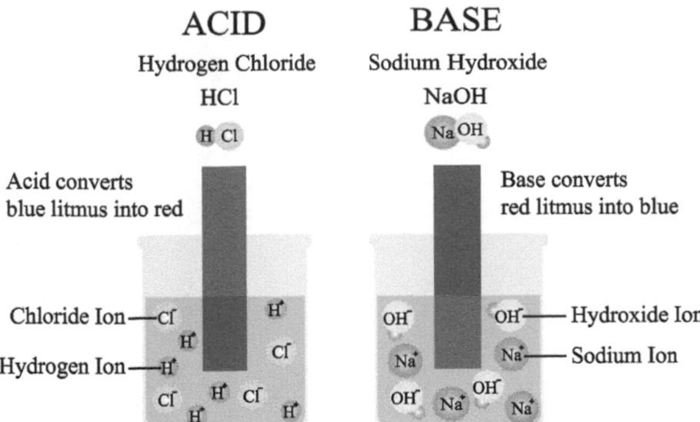

Litmus paper. Identifies a solution as acidic if it turns blue litmus paper red, or basic if it turns red litmus paper blue.

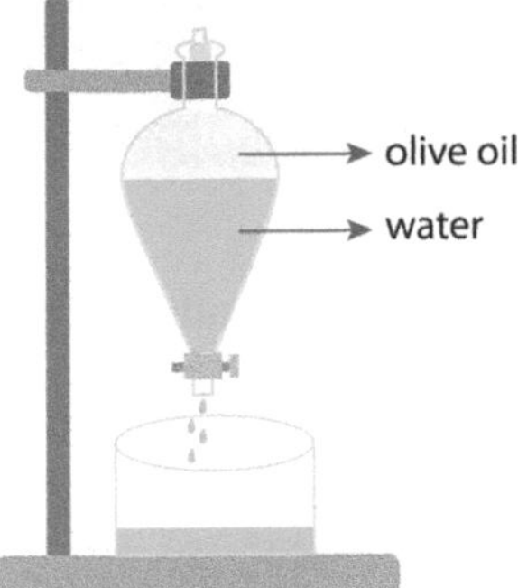

Separatory funnel. Allows the separation of immiscible liquids such as oil and water, by opening the stopcock at the base.

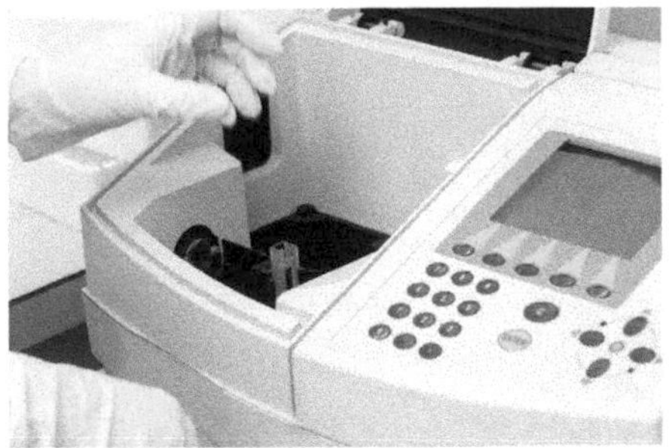

Spectrophotometer. Determines concentrations, identities, and properties of samples based on their interaction with light. Sample containers are called cuvettes.

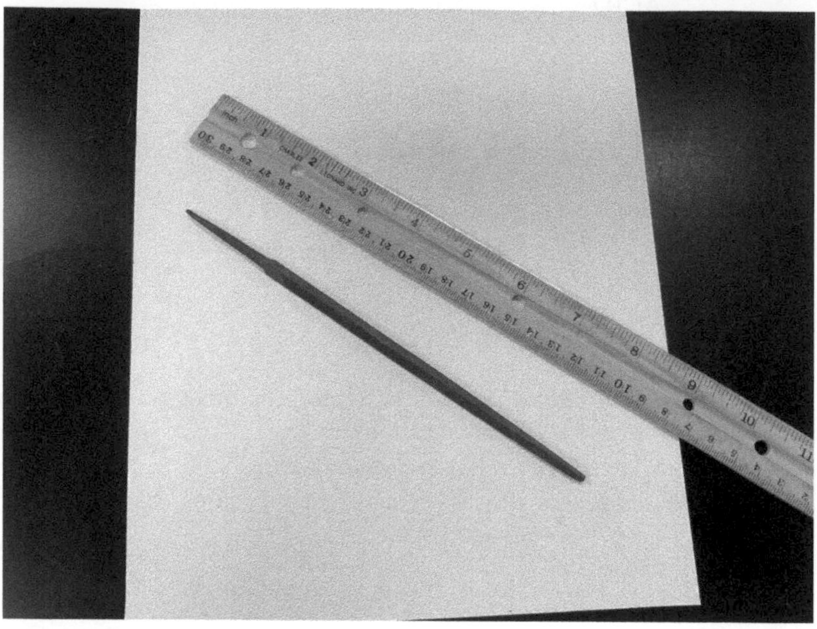

Triangular file. Useful for smoothing rough edges of glass tubing or metal samples.

Watch glass with chemical sample. (Erlenmeyer flasks in background.)

Weighing boat with sample. Although considered to be a disposable container for weighing out chemicals, it can often be washed and reused in the lab.

UNIT A

Introductory Exercises

EXERCISE 1

Safety in the Environmental Chemistry Lab

An ounce of prevention is worth a pound of cure. – Benjamin Franklin, American Scientist and Founding Father

Figure 1.1 Key lab safety reminders.

DOI: 10.1201/9781003006565-3

INTRODUCTION

Welcome to Environmental Chemistry Lab! A chemical laboratory provides a *wonderful* learning opportunity, and with careful attention to safety procedures and adequate supervision, you will find the chemical laboratory a safe learning environment. Figure 1.1 shows some basic safety reminders, and Figure 1.2 displays examples of typical glassware which you may use in lab. Your safety and the safety of your fellow students and instructor should be your prime consideration as you conduct laboratory experiments. If an experiment cannot be done safely, it is not done.

INSTRUCTIONS

First, study the following material. Next, complete the Safety Agreement and the Post-Lab Questions. Finally, keep this information in mind and at hand, as you work through various exercises in this text. Your instructor may ask you to turn in these pages at the end of your lab course.

The following should be observed when working in the chemical laboratory:

1. Give careful thought and planning to experiments.
2. Consider each step as an experiment progresses.
3. Take no chances.
4. Keep things cleaned up and orderly as you work.
5. Help others by calling to their attention unsafe practices.
6. See that your laboratory equipment is in good repair and safe operating condition.
7. Never sit on the lab benches – traces of chemicals could be present on them.
8. Above all, **Know What You Are Doing!** Read experiments before coming to lab, taking special note of potential hazards and indicated safety precautions. Attend the pre-lab lecture before entering the lab. This is important as safety is covered in these sessions.
9. Your entire desk area must be cleaned before you leave lab. Failing to do this can lower your grade.

Figure 1.2 Typical lab glassware.

Read the following lab safety policies, and then complete the Safety Agreement and answer the questions. Your instructor may discuss these with the class.

SAFETY GUIDE FOR THE ENVIRONMENTAL CHEMISTRY LAB

1. Approved safety goggles must be worn in the laboratory at all times. All Personal Protective Equipment (PPE) which consists of at the very least, approved splash goggles and lab apron or lab coat, must be stored when not in use, in closed, ziplock bags labeled with your name, if stored in the lab. Gloves are for single use and are disposed of after each use.
2. Anyone wearing contact lenses is required to notify her/his lab instructor before working in lab. (This is done on the Safety Agreement, which will be discussed later.) It is especially important for contact lens wearers to wear proper eye protection at all times. Your instructor may ask you to place a colored sticky dot on the right side of your splash goggles if you wear contact lenses in lab.
3. Rubber or plastic aprons – or approved lab coats – are required for all laboratory work.
4. Impermeable (non-canvas) footwear that completely covers the feet is required. (Leather footwear is not required.) Sandals and flip-flops are not acceptable.
5. Shorts, short skirts, short sleeves, halter tops, ties, long scarves, leggings, or fishnet shirts are not to be worn in the laboratory. Long sleeves are required – no sleeveless or short sleeves allowed. Crewneck tops are strongly recommended. Clothing should not be excessively tight or loose as either extreme is hazardous in the chemistry lab.
6. Know the locations of the fire extinguisher(s), fire blanket, eye wash station(s), safety shower, and first aid kit.
7. Notify your instructor immediately in the case of any accident such as injury, breakage of glassware, or chemical spill, especially on your clothes; also, inform the instructor of any event that nearly resulted in an accident.
8. Do not smoke in the laboratory.
9. Do NOT eat, drink, chew gum, or use tobacco in the laboratory. Dangerous chemicals may get in your mouth! Never taste anything in the laboratory. No chemicals are ever to be taken from the laboratory unless you are specifically told you can do so.
10. Horseplay and practical jokes are not permitted in the chemical laboratory.
11. All unknown chemicals, reagents, samples, etc. are to be considered hazardous until assurance to the contrary is obtained. Household chemicals in the lab such as salt, sugar, ice, etc. are to be treated with caution and never ingested as they may contain dangerous chemical contaminants. Read labels carefully and avoid problems such as shown in Figure 1.3
12. Follow closely the directions given in the laboratory manual, and those given by the laboratory instructor when attempting an experiment. Do not change the experiment without the instructor's authorization.

13. Do only the experiments assigned and in the manner prescribed. Unauthorized experiments are prohibited.
14. Proper care of equipment and working areas is an essential step in safety. A neat and orderly working area promotes safety.
15. Apparatus must be assembled in a stable and orderly fashion.
16. Apparatus and reagents, after use, should be returned promptly to their place of storage. Do not return chemicals to their original containers unless you are specifically instructed to do so.
17. Dispose of all dry chemicals, broken glass, waste paper, and scrap metal, in the appropriate waste containers. Dispose of all liquids and solutions as directed by your instructor. Do not pour any solutions down the sink unless your instructor has indicated that it is safe to do so. Matches, filter paper, paper towels, and other insoluble objects are NOT to be disposed of in the sinks!
18. Any buckets of sand in the laboratory are for putting out certain types of fires and controlling spills. They are NOT trash cans.
19. Spilled materials, whether liquid or solid, corrosive (capable of producing chemical burns) or inert, must be cleaned up promptly and completely. Consult your instructor for directions.
20. Cracked or chipped glassware must be immediately disposed of in the proper manner. Do not *toss* it in the receptacle as it may shatter; carefully place it in the designated container.
21. Fumes presenting fire or health hazards must be vented into a hood.
22. When testing the odor of a substance, always hold the container at a safe distance and waft the odor to your nose. Never stick the container right under your nose or inhale deeply.
23. Always point the mouth of a test tube away from yourself and anyone working near you when you heat its contents.
24. If using pipettes, mouth suction is never and must never be used. A pipette bulb or other device is needed.
25. Open flames should be avoided near flammable (able to catch fire) solvents.
26. Open flames are very hazardous and great care must be exercised with regard to baggy sleeves, sweaters, ties, and other objects that are flammable. Remember that hair is also very flammable. Confine or securely tie hair that reaches to your shoulders.
27. Soap and water, if frequently used on the hands and arms, will remove unsuspected bits of irritating chemicals and may serve to prevent painful cases of chemical dermatitis. Wash up after lab. Never apply Alconox™ (lab detergent) which is corrosive to the skin.
28. Acids as well as alkalis (bases) in contact with the skin will cause burns. Should either be spilled on the body, rinse promptly with large amounts of water.
29. Acids (especially concentrated sulfuric acid), when being mixed with water, must be poured into the water. NEVER the reverse! (Remember, "AA" – "Add Acid.").
30. Droplets of corrosive liquids on the side of reagent bottles after pouring must be removed before the bottle is returned to storage.
31. Stoppers from reagent bottles must not be placed where a person may lay her/his hand or arm on it, nor in such a way that the table top would be contaminated by reagent on the stopper or transfer impurities to the stopper.

32. Inserting glass tubing or thermometers into stoppers or rubber tubing is hazardous, and should be done with caution. To remove glass tubing or thermometers from rubber tubing or stoppers, the rubber tubing should, in most cases, be cut away from the glass.
33. Glassware under vacuum must be protected from physical shock that might lead to cracks, since a crack quite often leads to collapse with explosive violence.
34. Vacuum on any system must be relieved before any attempt is made to disassemble the equipment.
35. Never heat a closed container!
36. Obtain approval from your instructor before using any gas cylinders or compressed air. Pressures are high and valves must be opened slowly.
37. Hot silicone or water baths must be shielded and protected against breakage and spillage.
    These should not be mounted above eye level; do not place any chemical above the eye level of anyone in lab.
38. Volatile (easily evaporated) solvents when shaken in separatory funnels may develop considerable pressure. Rupture with explosive violence may be avoided by frequent removal of the stopper (or opening the stopcock when the funnel is inverted). Do not look directly into the funnel as liquid under pressure may shoot out.
39. Heaters, hot plates, etc., although not in evident use, may be hot. Never assume such equipment is cold. Remember, too, that hot glassware looks just like cold glassware. Use the same sort of caution with glassware as with hot plates.
40. You are responsible for reporting any allergies to chemicals, latex, etc. (excluding prescription drugs) in the space provided on the Safety Agreement.
41. Participation in chemistry lab requires handling of and/or exposure to potentially hazardous chemicals. A list of these chemicals can be found in each laboratory in the Safety Data Sheet (SDS) manual. If you are pregnant or have any medical condition which may be aggravated by exposure to these chemicals, it is strongly recommended that you consult with your healthcare provider prior to participation. If your physician determines that exposure to a chemical may be hazardous to your health, please contact your instructor to determine if you are eligible for an academic adjustment or other accommodation.
42. It is the student's responsibility to study each assigned lab prior to attending, and to inform the instructor *on the day of that lab*, of any chemical used in that day's lab, to which they might be allergic.
43. Lab desks, equipment checked out for the day, and equipment kept in the lab drawer, must be left clean. Dirty desk surfaces and equipment can be hazardous and difficult to clean if not maintained properly. When you leave the lab, you must inform the lab coordinator or instructor that you are leaving, and have your work area checked. Failure to follow this procedure can result in a lowered grade for the lab work.
44. Calculators, if used in lab, must be in sealed ziplock plastic bags. Calculators work well while in the bags. Cell phone usage is generally not allowed in the chemistry lab. Consult your instructor with any concerns.
45. You should immediately inform your instructor if spills of blood or body fluids occur; students are not to clean these up.

46. Gloves may not be worn outside of the lab and must be disposed of in the used-glove receptacle, not in the trash.
47. Know the hazards of any chemical before using it. The Globally Harmonized System (GHS) pictograms in Figure 1.4 can assist you with this. (This lab text strives to use the safest chemicals available. Some of these pictograms are for information only.) You may notice these pictograms on trucks hauling chemicals.
48. Beral (plastic) pipets must be disposed of in the receptacle labeled "Beral Pipets," not in the trash.
49. Weighing boats must be washed and returned to the supplies counter. Do not place these in the trash.
50. Read each label *three* times before using a chemical – some formulas are very similar, e.g., $NaOH$ and NH_4OH.
51. In the event of a fire drill, or fire, labs must be evacuated immediately, with students and personnel meeting at the site designated by the instructor.
52. No chemicals are to be placed in the trash. Used filter paper, pH paper, and litmus paper must be placed in the receptacle labeled "Used Filter Paper, pH Paper, and Litmus Paper." These are not placed in the trash.
53. The use of personal audio and visual equipment and cell phones is prohibited in the laboratory. These could pick up chemical contamination as well as be a distraction.
54. Students are responsible for the care and sanitation of their own PPE (splash goggles and lab apron or lab coat).
55. Your instructor may provide additional details or requirements as to lab safety.
56. A safe lab is not only an excellent learning environment, it provides enjoyable experiences in observing chemical processes.

SAFETY EQUIPMENT

A. Fire extinguishers – Carbon dioxide fire extinguishers are mounted in strategic locations throughout the laboratories.
B. Safety showers – Safety showers are provided for flooding a person with water in case of spillage of corrosive liquids on the body or clothing, or flames on the clothing.
C. Eye wash facilities
D. Fire blanket
E. First aid kit

Figure 1.3 Activities to avoid in lab! (Courtesy of Morgan Raines.)

IF IN DOUBT, CONSULT YOUR INSTRUCTOR!

Remember, by avoiding risks (Figure 1.3) and following lab safety instructions, you can be as safe in lab as you are at home – and possibly safer!

STUDENT ENVIRONMENTAL CHEMISTRY LAB SAFETY AGREEMENT

Basic rules have been defined on the Student Safety Information Sheet. The student reviews these basic rules and then signs this form, agreeing to abide by these rules and any additional safety directions provided by the science instructor.

The purpose of the agreement is to make the student aware of her/his responsibility for laboratory safety.

Students should also realize the implications of improper behavior. For example, courts have ruled that students can be just as guilty of negligence as teachers in laboratory accidents.

I Will:

- Follow all instructions given by the instructor.
- Protect my eyes by wearing approved safety goggles at all times in the laboratory.
- Protect my body (no shorts, halter tops, ties, long scarves, fishnet shirts, tight-fitting jeans, leggings, short sleeves, or sleeveless apparel should be worn in lab) by wearing my lab apron or lab coat (must be clean) at all times in the laboratory. Excessively tight or loose sleeves are not allowed in the lab.
- Protect my feet by wearing shoes that completely cover the foot. (Sandals and flip-flops are not acceptable.) Impermeable (non-canvas) shoes should be worn.
- Not smoke, eat, drink, or chew gum in the laboratory.
- Conduct myself in a responsible manner at all times (no horseplay of any type) in the laboratory.
- Not sit on the lab benches
- Carry out good housekeeping practices.
- Not work alone in the lab
- **Refrain from using cell phones, ear buds, or other distracting devices in the lab.**

I Understand:

That if I have, or develop during this semester, a special health concern (chronic or acute illness, pregnancy, etc.), I should consult with a physician before attending lab.

I, ______________________________ (printed name, **last name first**), have read and agree to follow the safety regulations set forth in this "agreement." I will closely follow the oral and written instructions provided by the instructor. I understand that failure to follow safety instructions can result in my dismissal from the lab with a grade of zero.

____________________	____________________	____________________
Student's Signature	Date	Course and Section

Contact lens wearer? Yes __ No __ (Check "Yes" if you will wear contact lenses in the laboratory at any time.) If yes, your instructor may provide a colored adhesive dot to be placed on your splash goggles, to the side where it will not interfere with vision, as a reminder in the event of an accident.

Please list any known allergies to chemicals, latex, etc. (excluding prescription drugs):

____

____

GHS HAZARD PICTOGRAMS

GHS01: Explosive **GHS02: Flammable** **GHS03: Oxidizing**

GHS04: Compressed Gas **GHS05: Corrosive** **GHS06: Toxic**

GHS07: Harmful **GHS08: Health Hazard** **GHS09: Environmental Hazard**

Figure 1.4 Globally Harmonized System (GHS) of classification and labeling of chemicals.

EXERCISE 1 SAFETY IN THE ENVIRONMENTAL CHEMISTRY LAB

Post-Lab Questions

Last Name____________________ **First Name**____________________
Instructor____________________ **Date**____________________

When answering multiple-choice questions, select the *best* answer.

1. What does PPE stand for?

2. Name the two main types of required PPE.

3. Shoes worn in lab should be
   a. made of impermeable material and cover the foot
   b. sandals to stay cool
   c. impermeable material
   d. closed-toe
4. Sand buckets in the lab are for
   a. trash
   b. decoration
   c. putting out fires
   d. putting out fires and controlling spills
5. Cracked or chipped glassware should be
   a. used carefully
   b. used only to contain safe chemicals such as soap and water
   c. used only for experiments that do not require heating
   d. discarded
6. Chemicals must be kept
   a. below the eye level of the instructor
   b. on a high shelf where they are out of the way
   c. below the eye level of the shortest person in the lab
   d. none of these

7. Alkali is another term for
   a. caustic
   b. corrosive
   c. basic
   d. radioactive
8. Flammable is a term meaning
   a. likely to catch fire
   b. not likely to catch fire
   c. likely to evaporate
   d. likely to sublime (change directly from solid to gas)
9. Volatile chemicals can easily
   a. catch fire
   b. evaporate
   c. solidify
   d. liquefy
10. In case of a chemical spill, you should first
    a. text your best friend
    b. evacuate the lab
    c. notify the instructor
    d. clean it up quickly before anyone notices
11. Chemicals which can burn the skin or eyes are called
    a. flammable
    b. volatile
    c. hazardous
    d. corrosive
12. In the event of an accident in lab, you should notify ____ as soon as possible.
    a. the instructor
    b. your roommate
    c. your lab partner
    d. the dean
13. Any special medical conditions pertaining to working in the lab should be approved by
    a. your friends
    b. your classmates
    c. your family
    d. your physician
14. Leggings are acceptable to wear in lab.
    a. True
    b. False
15. Shorts are acceptable attire for lab if a long lab apron is worn over them.
    a. True
    b. False
16. Ties and high heels are stylish and excellent attire for working in the lab.
    a. True
    b. False

17-25. Refer to Figure 1.1 and write a safety statement for each of the diagrams shown. List these in Table 1.1 A clearly written single sentence for each is acceptable.

Table 1.1 Post-Lab questions 17–25. Indicate in the designated column how the item below relates to lab safety. For example, if lab coat were listed, you might answer, "Protects worker from chemicals." If shorts were listed, you might answer, "Not allowed in lab."

Diagram	Your Answer – Relevant Safety Policy
Sandals	______________
Goggles	______________
Broken glass keeper	______________
Emergency eye wash station	______________
Food	______________
! pictogram	______________
Instructor	______________
Gloves	______________

EXERCISE **2**

Ethics in the Environmental Chemistry Lab

In matters of style, swim with the current; in matters of principle, stand like a rock.

– Thomas Jefferson, American Founding Father and President

Figure 2.1 Research and science ethics crossword, a reminder of the primacy of ethics in science.

INTRODUCTION

We would like to think that scientific research is always conducted ethically, as Figure 2.1 reminds us it should be; however, it is sad but true that some scientists have derailed their careers with ethics violations. These can range from falsifying data for lab reports in a science course, to falsifying data to cover up environmental infractions, and/or to gain a patent, promotion, or other advantage. There was even a case of a scientist presenting questionable work (which is now discredited), on psychoactive drugs for mentally retarded children. While the development of new pharmaceuticals can be profitable as Figure 2.2 indicates, the needs of the patient should be paramount. As

DOI: 10.1201/9781003006565-4

students preparing for a career, (1) you should know that this type of fraud exists to some extent even in the scientific community, and (2) you should avoid any temptation to present work that is not your own for grading, or to help someone else do this. Plagiarism is copying someone else's work without giving credit for it. And, yes, your instructors will remember any such infractions if asked to write a letter of recommendation for you. In the immortal words of Stan Lee, longtime Marvel Editor-in-Chief "Nuff said."

Figure 2.2 Ethics in Environmental Chemistry should involve all sectors of community life.

INSTRUCTIONS

Keeping in mind the high ethical standards for professionalism in science, complete the post-lab questions

ETHICS IN THE ENVIRONMENTAL CHEMISTRY LAB

Post-Lab Questions

Last Name ______________________ **First Name** ______________
Instructor ______________________ **Date** ______________________

Mark the following with a check-mark "√" for ethical lab work or an "X" for unethical lab work.

_____ 1. A student who does not want to perform an experiment on acid rain asks to copy another student's lab data.

_____ 2. A student who was absent from a lab on classifying plastics for recycling receives the instructor's permission to copy another student's lab data.

_____ 3. A student determines the density of copper with three trials, obtaining the results 8.9 g/mL, 8.2 g/mL, and 4.0 g/mL. The student changes the results in the notebook for the third trial to read 8.5 g/mL.

_____ 4. A student determines the density of iron, getting three results which agree closely and a fourth which is quite different. The student performs a standard deviation calculation, and, based on the statistical result, disregards the anomalous result.

_____ 5. A student is performing an electroplating experiment to see how metals are used in society. The instructions state to let the experiment run for 2 hours. Being hungry for lunch, the student leaves after 45 minutes and writes up the experiment as if it had continued for 2 hours.

_____ 6. A student performs a molecular weight determination on nitric oxide, an air pollutant generated by automobiles. While writing up the report later, the student notices that the results are poor, so he calculates backward to see what the data should have been, and then goes over the ballpoint pen entries with a broad, felt-tipped pen to make the results appear correct.

_____ 7. A student asks to copy another student's answers to post-lab questions on water analysis.

_____ 8. A student asks another student to explain the answers to one or more lab questions, and then writes the answers in her own words.

_____ 9. A student fills in information from the Internet for her lab report, rather than actually doing the experiment herself.

_____ 10. A student answers a lab question by copy-pasting from the lab text rather than putting it in his own words.

BIBLIOGRAPHY

Cronin, Brian, Comic book legends revealed #463, https://www.cbr.com > comic-book-legends-revealed-463

Garfield, Eugene and Welljams-Dorof, Alfred, Comic book legends revealed 463, *JAMA*, 1990, 263(10), 1424–1426. doi:10.1001/jama.1990.03440100144021

Jefferson, Thomas, *"The Jefferson Bible [annotated]" Original Old English Version and Modern Updates to The Jefferson Bible*, BookBaby, 2013, p. 139, https://www.azquotes.com

EXERCISE 3

The Laboratory Notebook in the Environmental Chemistry Lab

Put it in writing!

Figure 3.1 Entering results in a laboratory notebook with regular binding.

INTRODUCTION

Why do we say, "Put it in writing"? One reason is to keep an authentic record of a business transaction. Why do environmental chemists need a lab notebook like that shown in Figure 3.1? Lab notebooks can be invaluable! They can document findings which need correction, such as excessively high nitrate levels in natural

DOI: 10.1201/9781003006565-5

waters. They may contain information that is marketable, such as a new discovery. Lab notebooks, properly done, can stand up in court, and help you win in a patent dispute or other lawsuit. They can also help you pass science courses if your teacher requires that you keep a notebook. While you may not plan to be a scientist, an appreciation for the lab notebook will give you a better idea of how science works. Living in this scientific age, we all need an understanding of science. You should regard your lab notebook like your cell phone. Know how scary it can be to lose your phone? You don't want to lose your lab notebook, either! It can be the key to success in your science class, and to getting credit for a discovery in the job world.

INSTRUCTIONS

Here are some pointers for a successful lab notebook. Notice that the requirements are different from journaling, or writing a story – or a letter.

1. Write your name and course number on the notebook's cover.
2. The first page should include your name, the name of your course, the date, and your contact information (in case the notebook goes to the Lost and Found).
3. Begin a Table of Contents, and leave about 4 pages for future entries.
4. It is best to avoid using a spiral-bound notebook as shown in Figure 3.2. A

Figure 3.2 Spiral-bound lab notebook.

notebook with regular binding and pre-numbered pages is recommended. If your instructor allows you to use a spiral version, number the pages as you work your way through the notebook, and never remove a page. This is to avoid the appearance of removing results that seem erroneous.

5. Write the title and date of the experiment at the top of the page; also list the names of your lab partner(s) if you worked with someone else.
6. Write the procedure you followed. This can be shortened by just referencing a textbook process, but it should be clear enough that someone else could duplicate the procedure. Be sure to include any changes from the textbook procedure. Examples are: (1) omitted section B on pipette calibration, or (2) substituted a hot plate for a Bunsen burner. It is a great idea to list the steps you will follow in the lab; lab textbooks are all too often written with procedures in somewhat of a random order. For example, you may need boiling water at some point in the experiment, but the lab text may suddenly just say to drop a piece of copper wire in boiling water. So, first read the lab text, and write in your notebook the procedure in the order the steps should occur. This can save you a lot of time in the lab, e.g., knowing you will need boiling water, you can start heating it in advance.
7. Write in blue or black, non-erasable ink. Do not use a felt-tipped pen, pencil, or a glitter pen.
8. If you need to change an entry, never erase or totally scratch out the first entry. Just draw one line through it and write the corrected information. Never use white-out, either. You do not want to give any appearance that the data that you collected in lab were later changed to make the results more impressive, nor do you want your work to appear less than professional.
9. Write neatly so that your work is legible and easily read.
10. Think tables, rather than long paragraphs. Imagine how much reading would be involved if the table in Figure 3.3 were presented as a collection of sentences. Notice how listing the units in the headings saves the bother of listing them after each entry.

ANALYSIS OF WATER FROM LAKE LOTTA-RAIN

DATE OF COLLECTION	NITRATE ION CONCENTRATION mg/L (ppm)	NITRITE ION CONCENTRATION mg/L (ppm)	SULFATE ION CONCENTRATION (ppm)
01/05/2022	10	0.1	29
01/12/2022	10	9.8	30
01/19/2022	15	1.0	25
01/26/2022	8	0.5	31

Figure 3.3 Sample data table.

11. Never leave excess blank space on a page. If your entries do not fill the page, sign your name at the end of the information you wrote, and make a big "X" on the remainder of the page. Your instructor may want to look over your data and sign the page too. This verifies that nothing was added after you finished the experiment and left the lab. Surprisingly, even highly respected scientists have been known to fabricate data for personal advancement.
12. Protect your notebook from laboratory spills, heat sources, etc.
13. Remember that your notebook contains the data you yourself collected for the

experiment. Never share this information with other students, unless your instructor approves.

14. Try to make your notebook look as neat and professional as possible. When entering a data table such as shown in Figure 3.3, e.g., using a straight edge only takes a few extra seconds and makes the document look much better than just drawing the lines by hand.
15. Be sure to describe your results clearly. Recording anything unusual that happened is not superfluous – it is good science.
16. A lab notebook should be the *original* record of your work, so never record information on scraps of paper or other places to be transferred later to your notebook. Sometimes students say that they do not record the data on their notebook at first, because they want to wait until they can write it very neatly later at home; however, this is not the best practice. Record your information in your notebook – never on other paper and never on your hand.
17. Never record data from two experiments on the same page; even if you have space left, remember to draw a big "X" on the space, and begin the next experiment's records on a fresh page.
18. Any abbreviations used in your notebook should be clearly defined, unless they are well known to the scientific community, such as mL for milliliters. And, unlike in English class, for scientific writing we do not use a period after an abbreviation. In other words, kg for kilograms rather than kg. is required.
19. While there is some lab work done currently with electronic notebooks as shown in Figure 3.4, many of the principles covered here refer to that format as well, and rest assured that the traditional lab notebook described above is "alive and well!"

Figure 3.4 Scientist working in lab with a tablet computer.

EXERCISE 3 THE NOTEBOOK IN THE ENVIRONMENTAL CHEMISTRY LAB

Post-Lab Questions

Last Name ____________________ **First Name** ____________________
Instructor ____________________ **Date** ____________________

(60) 1. List at least six ways a laboratory notebook differs from a letter you would write to a friend.

(10) 2. How could failure to record data properly

a. cause someone to lose a valuable patent for an invention?

b. contribute to an environmental crisis?

(10) 3. Why are felt-tipped pens not allowed for recording data in lab notebooks?

(10) 4. Explain why lab notebooks have numbered pages. (Two reasons are given in the above discussion.)

(10) 5. Explain why any abbreviation you use in your notebook should be clearly identified, unless it is a well-known scientific abbreviation, such as °F for degrees Fahrenheit.

EXERCISE 4

Introduction to Glassware in the Environmental Chemistry Lab

Now we see through a glass darkly.

– I Corinthians 13:12 King James Version

Figure 4.1 Glass tableware, like labware, can be extremely fragile and costly.

INTRODUCTION

Environmental chemists appreciate the importance of glassware and keeping it clean! The above quotation would not be very appropriate for an Environmental Chemistry Lab. We expect tableware like that shown in Figure 4.1 to be clean, and, if we want good experimental results, lab glassware is no less important.

DOI: 10.1201/9781003006565-6

INSTRUCTIONS

Your instructor may present the following material or assign it for you to read. After you have studied it, complete the worksheet.

OVERVIEW

In the lab, contaminants can cause erroneous results, such that pollution is not identified and therefore not cleaned up. Other labs in this text use glassware to measure air quality, water quality, and more. Glassware is a key component of a science lab for both qualitative and quantitative studies. *Qualitative studies* in Environmental Chemistry tell us *what* contaminants, such as the nitrate ion and the nitrite ion, are present, while *quantitative studies* tell us the amounts of these. The US Environmental Protection Agency (EPA) has established 10 mg/L of nitrate ion and 1 mg/L of nitrite ion, as the **Maximum Contaminant Level (MCL)** considered to be safe. So, both types of studies are extremely important. Glassware can be quite costly, and its proper care is vital to a good lab experience. Broken, chipped, or cracked glassware is a safety hazard. Dirty glassware is esthetically unappealing as well as unreliable or even hazardous. What is glass, anyway? Some call it an **amorphous solid** – it *seems* solid, but the molecular structure is messy and chaotic, rather than organized neatly like the sodium chloride structure (Figures 4.2 and 4.3).

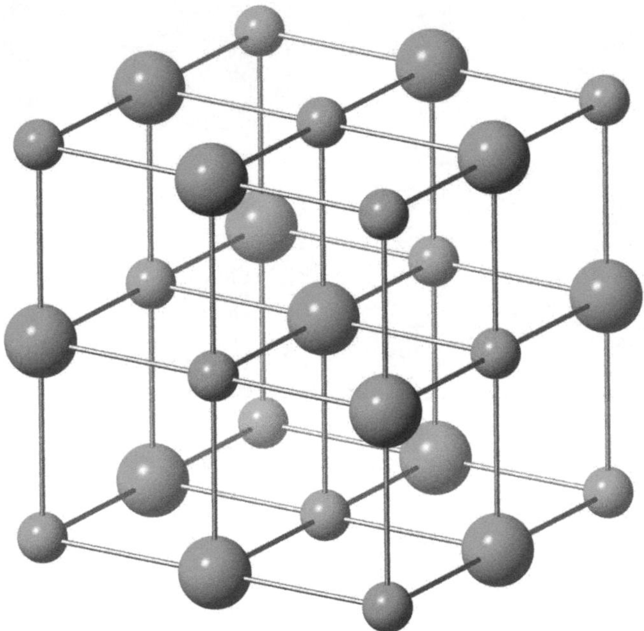

Figure 4.2 Sodium chloride (salt) structure.

Figure 4.3 Common salt shaker.

Glass is commonly made from **silica** (aka sand or silicon dioxide), sodium carbonate, and calcium and/or magnesium carbonate. Coloring agents are sometimes added. Cobalt oxide yields a blue color, chromic oxide green, etc. Uranium (a nuclear weapons element) oxide has even been used for stained glass windows in churches. Chemistry is truly amoral – the user of chemicals, not the chemicals themselves, determines if the chemicals help or hurt people!

Another interesting additive to glass is **boron** (Figure 4.4), a **metalloid.**

Metalloids are elements located somewhat centrally in the periodic table, and they resemble both metals and non-metals. The addition of boron to glass yields **borosilicates** which produce the heat-resistant product Pyrex or its equivalent. Not only chemists but chefs, and others depend on Pyrex for its ability to withstand high temperatures without breaking. Always be sure your test tube is Pyrex or the equivalent before heating it!

And, then, there's lead (Figure 4.5). Yes, lead, a **neurotoxin** (which attacks the nervous system), is added to some crystal – it is beautiful and heavy, but lead can leach out of it into beverages, so it should be used cautiously.

PROTECTING GLASSWARE FROM SCRATCHES

Brushes and detergents are great for cleaning glassware, but if bristles are missing from the brush, the glass can get scratched instead of cleaned. Brushes should have intact, complete sets of bristles. Stirring rods should have rubber policeman tips, to avoid scratches.

Figure 4.4 Boron (B), element 5 of the periodic table.

Figure 4.5 Lead (Pb), element 82 of the periodic table.

PROTECTING GLASSWARE FROM HEAT

Always start heating your glassware with a cool (not a preheated) hot plate. Raise the temperature of the hot plate gradually, even if you are in a hurry. A calibrated hot plate is useful here. When evaporating a liquid, never heat the beaker until it is dry; leave a bit of solution in the container and remove it from the heat source. The residual heat will dry the beaker. If heating glassware with a burner, a wire gauze with sintered

Figure 4.6 Erlenmeyer flask with contents, being heated over Bunsen burner flame. The flask is supported by the ring stand and protected by wire gauze with a sintered glass center. The thermometer, which should not touch the glass, helps monitor the temperature.

glass center (made with glass *particles*) should be placed on the ring of the ring stand (Figure 4.6) between the glassware and the burner. Only Pyrex glass or the equivalent should be heated – never soft glass. See Part D, question 2 with image.

PROTECTING GLASSWARE FROM BREAKAGE

Beakers, especially the larger ones, should be lifted by the sides, never by the lip. Grasp larger beakers (1 L or more in volume) with *both* hands. Never add liquid to hot glassware.

If using round bottom flasks, secure them with a clamp or by placing them in a cork ring (Figure 4.7); otherwise, they may roll off the bench – like a bowling ball!

PROTECTING YOURSELF FROM BURNS

Hot glassware should be allowed to cool before it is picked up; otherwise, wear heat-resistant gloves or use tongs.

LEARNING THE ACCURACY OF GLASSWARE

You can estimate one more place than the graduations (lines) show on a graduated cylinder or other graduated labware. For example, Figure 4.8 shows a

Figure 4.7 Round bottom flask protected from rolling by cork ring.

100 mL graduated cylinder. Since it is marked in milliliters, you should estimate the volume to the nearest *tenth* of a milliliter; thus, the cylinder in the figure contains about 60.8 mL. Graduated cylinders have about 1% accuracy. So, if the liquid level in the cylinder is at 100 mL, to be as accurate as possible, we would say it contains between 99 and 101 mL.

Class A glassware is more accurate (and more expensive!) than Class B or unclassified glassware.

Erlenmeyer flasks (Figure 4.9) are named after the German chemist Emil Erlenmeyer who invented them. These and beakers have graduations for *approximations* – never exact measurements. They are about 5% accurate. In other words, a reading of 100 mL on an Erlenmeyer flask or a beaker can actually mean somewhere between 95 mL and 105 mL!

DETERMINING POSSIBLE ERROR

The relative uncertainty is the absolute value of the error, divided by the measured value. For example, if a 125 mL Erlenmeyer flask has a ±5 mL error, and we are measuring 125 mL, the relative uncertainty = 5 mL/125 mL = 0.04. To get the percent relative error, we would multiply the 0.04 by 100, and get 4% relative uncertainty.

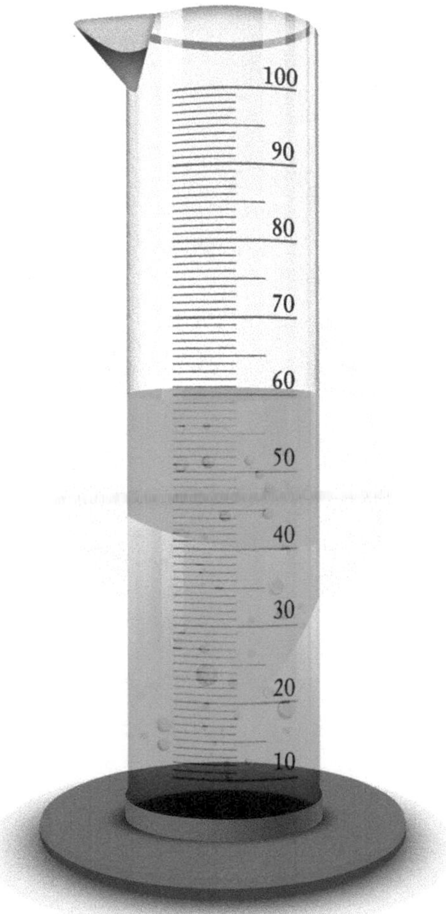

Figure 4.8 100 mL graduated cylinder.

Sometimes in Environmental Chemistry, we need to measure extremely small concentrations or calculate extremely small errors. In these cases, we may use parts per million (ppm) or even parts per billion (ppb). So, a 0.0006 relative uncertainty would be $0.0006 \times 1{,}000{,}000 = 600$ ppm relative uncertainty.

As another example, the US EPA has set an MCL of 2 ppm for mercury (another neurotoxin) in drinking water. Mercury in drinking water is sometimes reported in the ppb range (e.g., 17 ppb). This means the concentration was multiplied by 1,000,000,000. This is common and considered safe (although possibly unappetizing) for mercury.

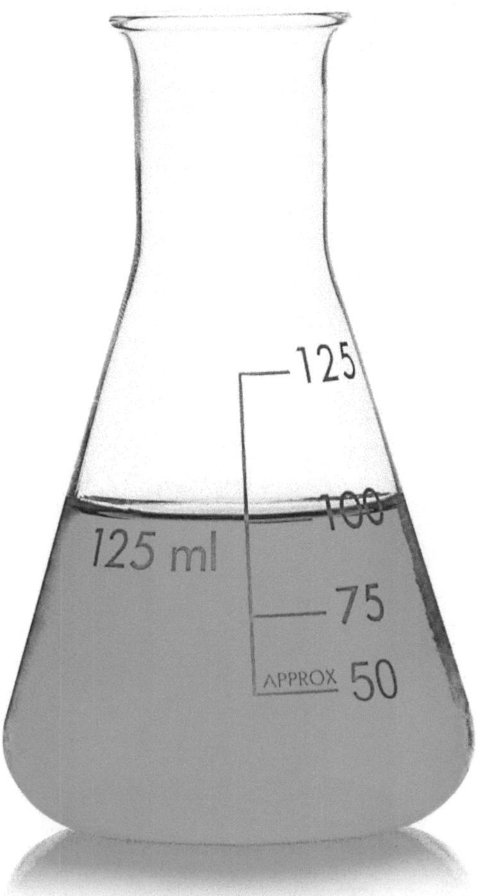

Figure 4.9 Graduated 125 mL Erlenmeyer flask.

WORKING WITH VOLUMETRIC GLASSWARE

Here we discuss some of the more expensive options. To prepare a solution in a volumetric flask such as those shown in Figure 4.10, you can add the **solute** (what gets dissolved) to the flask. Then add sufficient **solvent** (what dissolves the solute) to approximately half-fill the bulb of the flask. Put the cap on the flask tightly and shake the flask until the solute is all dissolved and mixed. Then add solvent until the top of the solution reaches the fiduciary line. Replace the cap and shake the flask until the solution is well mixed.

Now, transfer the solution to a storage bottle or other container. Never store solutions in volumetric flasks, as they can etch the glass, changing the volume. Storing **alkaline** (basic, i.e., pH more than 7) solutions, including alkaline detergent, in

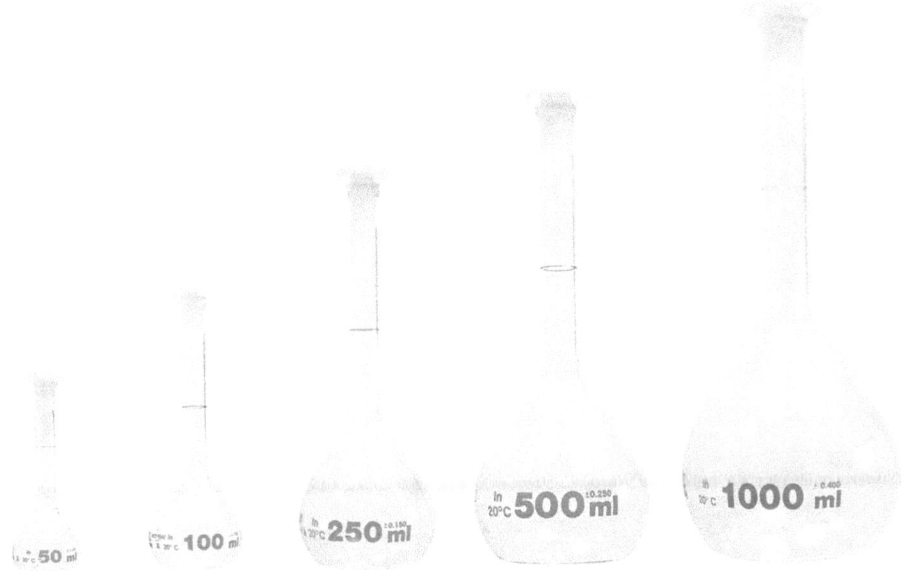

Figure 4.10 Volumetric flasks of various sizes.

volumetric glassware can be especially destructive to their accuracy. Similarly, do not heat volumetric glassware. This too can alter the volume.

Never mix chemicals like sulfuric acid with water in a narrow-necked container like a volumetric flask. The **Heat of Solution** generated in this process cannot all escape through the narrow opening and may burst the flask, spraying acid and glass on everyone around. Sulfuric acid is one of the chemicals having a very high heat of solution.

A **buret** (Figure 4.11) dispenses highly accurate amounts of liquid in variable amounts by control of a stopcock. These are often graduated in tenths of a milliliter, which means the liquid level in them can be estimated to the nearest hundredth of a milliliter. The general principle again, is to read graduated glassware to one more place than is marked. While the last place in the reading may not be exact due to estimation, it is useful. And it is a "rule of the road" that the last figure (on the far right) of a measurement is considered to be approximate. In other words, if a balance indicates a mass of 31.239 mg, the 9 which is the last digit, is considered an approximation. [If you ever have to use a balance where the numbers keep changing back and forth, you will experience this example].

Pipettes (Figure 4.12) are another means of transferring measured volumes of liquids between containers. Those labeled TC are "to contain" the given volume, so all of the contents should be transferred. In contrast, pipettes labeled TD are "to deliver" the given amount, so totally emptying the pipette would add too much of the liquid. Never blow the last remaining bit out of a TD pipette.

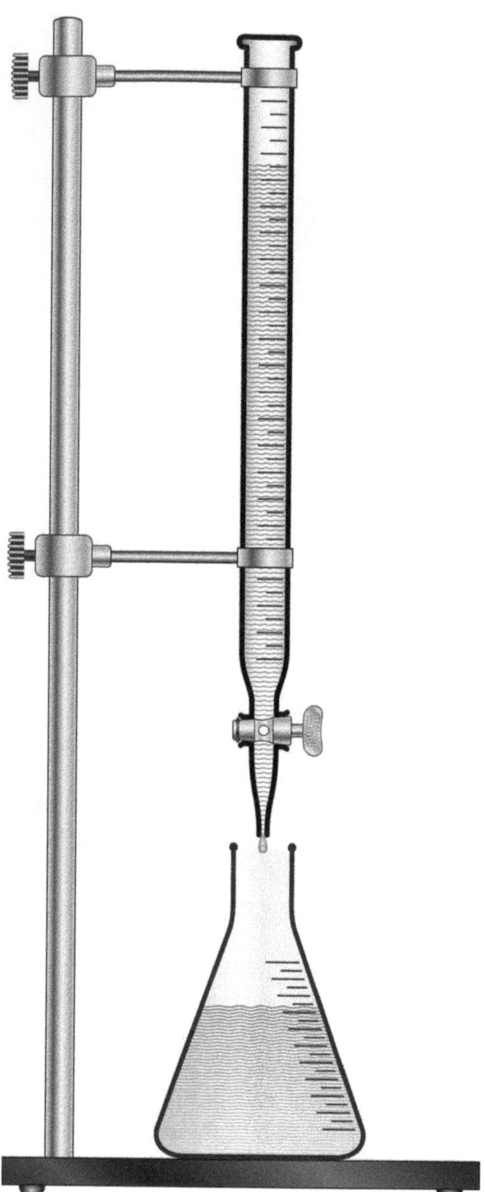

Figure 4.11 Buret dispenses solution into a graduated Erlenmeyer flask.

CLEANING GLASSWARE

As laborious as it sounds, for the best results, glassware should be washed before it is used, even if new, and rinsed with **deionized (DI) water** or **distilled water**. Glassware should also be washed after it is used, to avoid chemical residues

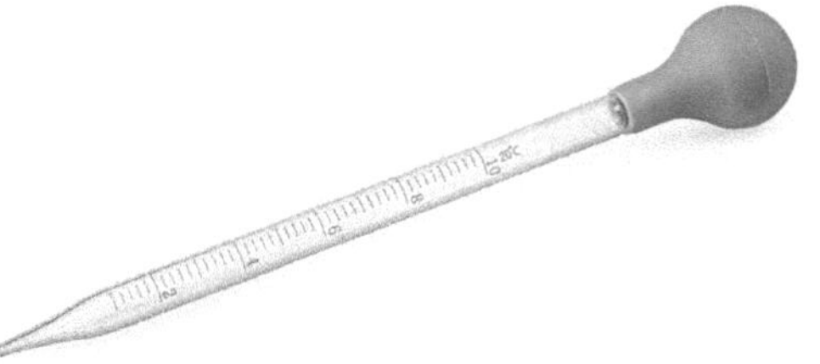

Figure 4.12 Pipette with bulb.

which would be difficult to remove. *Seven* DI rinses are considered sufficient for glassware used in research. And if you see water forming "beads" on the surface, it's time to scrub the glassware again – water will uniformly wet clean glassware. After it is washed, glassware can be rinsed initially with tap water, and then with deionized or distilled water. Traces of detergent should be removed – especially if adding acid to the container, as a salt appearing as a scum can form. Stubborn greasy residues can be removed with cautious use of **acetone** (CH_3COCH_3), a common ingredient in nail polish remover, or 10% aqueous sodium carbonate solution.

Now that our glassware is clean and rinsed, how do we dry it? Allowing it to dry on a rack is best; towels can transfer lint and compressed air can transfer oils, dirt, or even moisture. A word to the wise is don't put wet glassware away in a cabinet or drawer – unless you want to invite mold (Figure 4.13)! Actually, the story is that **Sir Alexander Fleming**, Scottish microbiologist and physician, discovered the powerful antibiotic penicillin due to mold growing on his glassware, but that was another time and another century.

Finally, last but not least, glass is fragile. Handle it with care.

Figure 4.13 Microscopic image of growing molds or mold fungus and spores – 3D illustration.

EXERCISE 4 INTRODUCTION TO GLASSWARE IN THE ENVIRONMENTAL CHEMISTRY LAB

Last Name ____________________ **First Name** ______________________
Instructor ____________________ **Date** ___________________________

Part I. Getting Acquainted with Laboratory Glassware

1a. Before washing beakers, always inspect beaker brushes for ____________
____

1b. Explain your answer to 1a. ______________________________________
____

2. To avoid scratching glassware, use only stirring rods with ______________
____

3. Always use a ____________________ between an open flame and the beaker bottom.
4. When using hot plates, place beakers on a
   a. cool surface
   b. preheated surface
5. When using hot plates, raise temperature
   a. gradually
   b. rapidly
6. Avoid heating beakers to dryness.
   a. True
   b. False
7. Lift beakers by firmly grasping the beaker's
   a. lip
   b. sidewalls
8. Use *two* hands to lift large (1000 mL or greater) beakers.
   a. True
   b. False
9. Cracked or chipped beakers should be
   a. discarded
   b. saved
   c. reserved for light use
   d. reserved for indoor plants
10. It is sometimes acceptable to add liquid to hot glassware.
   a. True
   b. False
11. How can hot glassware be picked up? Check all that apply.
   a. With heat-resistant gloves
   b. With tongs
   c. With plastic gloves
   d. None of these

Part II. Measuring with Laboratory Glassware

1. A graduated cylinder has about a ______________% accuracy.
2. Graduations on Erlenmeyer flasks and beakers have about a ____________% accuracy and should be considered
   a. exact
   b. approximate.
3. Calculate the relative uncertainty in ppm for a 500 mL ± 0.2 mL volumetric flask. Show method.

4a. A buret graduated in *tenths* of milliliters should be read to the nearest _________________ of a milliliter.

4b. A thermometer graduated in degrees (Celsius) should be read to the nearest _________________ of a degree.

4c. A graduated cylinder with marking for each milliliter should be read to the nearest _________________ of a milliliter.

4d. As a general rule, glassware volumes should always be read to _________________ place(s) than the graduations indicate.
   a. one more
   b. two more
   c. three more
   d. no more

5. In preparing a solution in a volumetric flask, the solute should first be dissolved completely in
   a. more solvent than is needed for the final volume
   b. less solvent than is needed for the final volume
   c. the exact amount of solvent that is needed to reach the mark on the flask.
6. Volumetric glassware may conveniently be used for storing solutions.
   a. True
   b. False
7. Explain your answer to question 6.

8. Explain why solutions such as those of sulfuric acid or others with highly exothermic heats of solution should NOT be prepared in a volumetric flask or other narrow-necked container.

Part III. Cleaning Laboratory Glassware

1. Glassware should be washed
   a. before using
   b. after using
   c. a and b
   d. Neither a nor b
2. Initial rinsing of glassware that has been washed should be with
   a. distilled water
   b. deionized water
   c. tap water
   d. none of these
3. Final rinsing of glassware that has been washed should be with
   a. distilled water
   b. deionized water
   c. tap water
   d. a or b
4. Glassware is not clean if water forms beads on its surface.
   a. True
   b. False
5. Glassware must be thoroughly rinsed to remove detergent residues before contact with any acid (even a weak acid); otherwise____________________
   ____
6. Grease can be removed from glassware by
   a. soaking in a 10% sodium carbonate solution
   b. cautiously using acetone
   c. a and b
   d. neither a nor b
7. Glassware should be dried by
   a. leaving it on a rack
   b. drying it with a towel
   c. using an air blower
   d. none of these
8. Volumetric glassware used for research purposes, after washing and rinsing with tap water, should be drained, and then rinsed ______________ times with distilled/deionized water.

9. ______________________ flasks should be secured so they do not roll off the lab bench and break.
10. Volumetric glassware should never be dried by ______________________
11. All glassware is
    a. fragile
    b. impact-resistant

Part IV. Identifying Types of Laboratory Glassware

1. Borosilicate (e.g., Pyrex^{TM}) glassware
    a. is resistant to large temperature changes
    b. is resistant to chemicals
    c. is more expensive than soda-lime (flint) glass
    d. all of the above
2. Predict what would happen if this battery jar made of soda-lime glass aka soft glass (not borosilicate) of water is heated.

3. Explain the difference between TC and TD pipettes.

4. Explain the difference between Class A, Class B, and unclassified volumetric glassware.

5. Volumetric glassware may be allowed to soak indefinitely in strong alkaline cleanser.
   a. True
   b. False
6. Explain your answer to question 5.

7. Explain why borosilicate glass, e.g., Pyrex, instead of regular glass, is recommended for work involving sudden temperature changes.

BIBLIOGRAPHY

geology.com/articles/color-in-glass.shtml

www.epa.gov/ground-water-and-drinking-water/national-primary-drinking-water-regulations#inorganic

EXERCISE 5

Exploring the Metric System

I often say that when you can measure what you are speaking about, and express it in numbers, you know something about it.

– Sir William Thomson (Lord Kelvin)

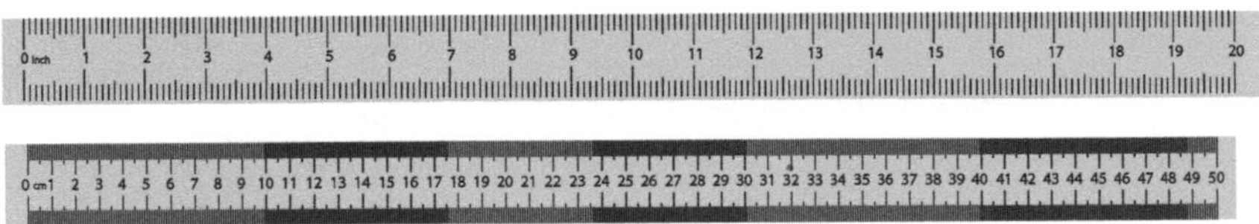

Figure 5.1 "Half-meter sticks" compare English units (inches) with metric (centimeters and other) units.

INTRODUCTION

Measurements are extremely important in Environmental Chemistry. We must know the concentrations of the chemicals in the air we breathe, in the water we drink, in the soil around us, and more! Invented by the French chemist Antoine Laurent Lavoisier, the metric system is renowned for its ease of use; units are related to each other by factors of ten. The United States is one of only three countries in the world which have not officially adopted the metric system. We do use the metric system – as you know, some beverages are sold by the liter, athletes run 5 kilometers, km, and many products are labeled with both English and metric units.

While the organization of the metric system is based on 10, e.g., 10 decigrams = 1 gram, 10 centigrams = 1 decigram, etc., the English system came about more randomly. An inch (2.54 centimeters or cm) was the width of a man's thumb, the yard (91.44 cm) was the distance from the King's nose to his outstretched hand, and a fathom (1.8 m) was a Viking's embrace!

DOI: 10.1201/9781003006565-7

INSTRUCTIONS

First, read through and practice the examples. Next, work the problems, following the method of dimensional analysis which is used in the solved examples.

Let's practice some unit conversions.

EXAMPLE 1 HERE'S AN EXAMPLE FOR YOU TO PATTERN YOUR SOLUTIONS BY.

Matcha Mole drives her green truck to the gasoline station and purchases 12.5 gallons of gas. Convert this to liters.

Step 1. Write down what you are given: 12.5 gal
Step 2. Find the needed conversion: 1.00 gal = 3.79 L
Step 3. Multiply the 12.5 gal by the conversion so that the given units (gal) cancel and the new units (L) remain:

12.5 ~~gal~~ × 3.79 L/1.00 ~~gal~~ = 47.4 L

EXAMPLE 2 SOMETIMES THESE CONVERSIONS CAN INVOLVE TWO OR MORE STEPS. FOR EXAMPLE, YOU MIGHT NEED TO CHANGE GALLONS TO QUARTS, AND THEN QUARTS TO LITERS, AS IN THE FOLLOWING PROBLEM.

Over a 10-year period, the 3M Company of St. Paul, Minnesota, reduced the quantity of wastewater produced by 1 billion or 1,000,000,000 gallons. Convert to liters, using the following:

1.0 gal = 4.00 qt and 1.00 L = 1.09 qt

Step 1. Write down what you are given: 1,000,000,000 gal
Step 2. Find the needed conversions: 1.0 gal = 4.00 qt and 1.00 L = 1.09 qt
Step 3. Multiply the 1,000,000,000 gallons by the conversions so that the given units (gal) and the intermediate units (qt) cancel and the new units (L) remain.

1,000,000,000 ~~gal~~ × 4 ~~qt/gal~~ × 1.09 L/~~qt~~ = 4,360,000,000 L

Now, try solving these. Remember to show your method, units, and how they cancel.

1. In 1 minute, the average human will breathe 7.0 quarts of air. How many liters is that?

   Hint: 1.00 L = 1.09 qt

2. Convert your answer from Question 1, to milliliters.

   Hint: 1000 mL = 1 L

3. Sneezes can travel 100 miles per hour! How many kilometers per hour would that be?

   Hint: 1.6 km = 1 mi

4. The average human eats about one ton of food per year. How many grams of food is that?

Hint: 1 ton = 2000 pounds and 1 pound = 454 grams

5. Convert your answer to Question 4 to kilograms (kg).

Hint: 1 kg = 1000 g

6. A human hair is about 80 microns (μ) wide. Convert to inches.

Hint: 1 in = 2.54 cm and 1 μ = 0.000001 m or 10^{-6} m

7. With their stretchy protein molecules called resilin, fleas can jump a height of 91.44 cm. Express in feet.

8. If a human had the same jumping capability as the flea, the human could jump a height of 16916.4 cm. Express in feet.

9. In June of 1985, the failure to make a 30-second calculation caused the space shuttle Discovery to go into a spin. The test of the "Star Wars" laser system consequently failed, though equipment functioned properly. Convert 12,000 feet to nautical miles and do what 3 or 4 Mission Control engineers overlooked.

Hint: 1.00 nautical mile = 1.15 miles

10. The Earth's water cycle includes total annual precipitation of 111,000 cubic kilometers (km^3). Convert to gallons.

$$\text{Hint: } 1\ \text{km}^3 = 264{,}000{,}000{,}000 \quad \text{gal}$$

BIBLIOGRAPHY

Beard, James M. and Murphy, Ruth Ann, *Environmental Chemistry in Society*, Third Edition, CRC Press, Taylor and Francis Group, Boca Raton, FL, 2022, p. 189.

Nibler, Joseph W. et al., *Experiments in Physical Chemistry*, Ninth Edition, McGraw Hill, USA, 2014, p. viii.

EXERCISE **6**

Mastering the Metric System

Antoine Laurent Lavoisier, the French inventor of the metric system, and his wife Marie Lavoisier are considered to be the Father and Mother of Modern Chemistry.

Figure 6.1 The Statue of Liberty. A gift to the United States from France, with dimensions stated in the metric system.

DOI: 10.1201/9781003006565-8

INTRODUCTION

We learned in the previous Metric System worksheet lab that the metric system, now called the SI for International System of Units, was invented by the Father of Modern Chemistry, Antoine Lavoisier. As he was assisted in the lab by his wife Marie, could she have helped him create this? She could be called the Mother of Modern Chemistry! We do know that when the Statue of Liberty was given to the United States by France, it was described in metric units, although some Americans complained that *American* units would have been preferred. They would likely be surprised to know that today this treasured symbol is subject to corrosion and other damage from environmental factors such as acid rain (Figure 6.1).

Given the Metric System's importance in today's environment, let's practice with some more problems.

Density is at the *heart* of chemistry; density = mass/volume, or d = m/v. Density helps explain oil spills; oil is less dense than water, so oil floats *on top of* the sea. Plastics can be classified for recycling based on their densities. We can also write:

$$\text{volume} = \text{mass/density}\,(v = m/d)$$

and

$$\text{mass} = \text{density} \times \text{volume}\,(m = dv).$$

INSTRUCTIONS

Show method, units, and how they cancel. You may wish to refer to the introductory material in the Exploring the Metric System Lab Exercise 5, to review the practice examples.

A. **Length**

1a. Convert 6 ft. 2 in. to centimeters (cm).

1b. Convert your answer from 1a to millimeters (mm).

B. **Volume**

2a. A water bottle contains 0.75 quarts of water. Convert this volume to liters (L).

2b. Convert your answer from 2a to milliliters (mL).

C. **Mass**

3a. The Statue of Liberty, which stands in New York Harbor, weighs 225 tons, or 450,000 pounds. Convert to grams. Hint: 1 pound = 454 grams

3b. Convert to kilograms (kg). 1 kg = 1000 g

3c. Convert the weight of a basketball of 1.25 pounds to grams.

D. **Density**

4a. Large sheets of copper (Cu) were used to cover the Statue of Liberty. Find the mass of the copper if the volume of the copper is 314,000 mL. Hint: Copper's density is 8.96 g/mL.

4b. Convert the 314,000 mL of copper to liters (L).

4c. Calculate the volume occupied by 309 g of osmium (the densest element known). Density of osmium = 22.5 g/mL.

EXERCISE 7

Parts Per Million (ppm) and Other Tiny Quantities

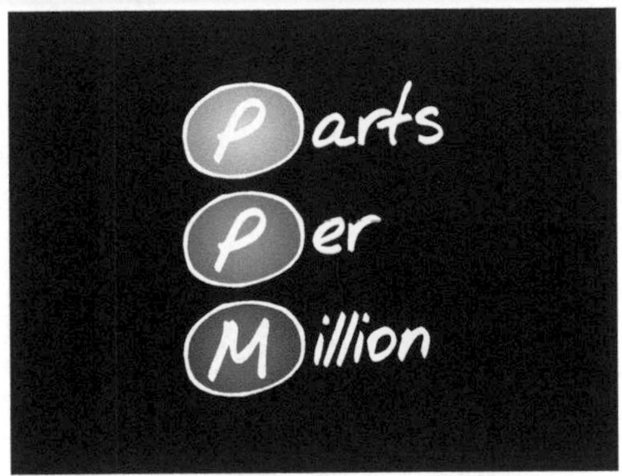

Figure 7.1 Parts per million (ppm) are common units for the measurement of environmental pollutants.

INTRODUCTION

Environmental Science sometimes deals with very tiny concentrations (Figure 7.1). As an example, the US Environmental Protection Agency (EPA) limit for the toxic metal lead in water is 15 parts per billion (ppb). This is the same as 15 micrograms per liter. A microgram is a millionth of a gram. The units ppb and parts per million (ppm) are convenient ways to express very tiny concentrations. You can think of them as "mini-percentages." Instead of multiplying by 100 to calculate the percentage, we multiply by one million (1,000,000) to calculate ppm and by one billion (1,000,000,000) to calculate ppb. See Table 7.1 for a conversion chart to help. Note that the

DOI: 10.1201/9781003006565-9

smaller the allowed amount, the more toxic the substance is, e.g., a substance with a Maximum Contaminant Level (MCL) of 1.2 ppb is 1000 times more toxic than a substance with an MCL of 1.2 ppm. And a substance with an MCL of 2 parts per thousand (ppt) is 5 times more toxic than a substance with an MCL of 10 ppt. Everything has to go somewhere, so if batteries containing lead, cadmium, et al, are discarded in the trash, their components can end up in the soil, waterways, etc. Proper recycling of batteries (Figure 7.2) is a far better choice.

Table 7.1 Concentration Units and Their Definitions

Term	Meaning
Parts per thousand	Grams/liter (g/L)
Parts per million	Milligrams/liter (mg/L)
Parts per billion	Micrograms/liter (μg/L)

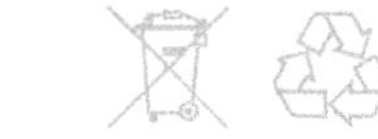

PLEASE RECYCLE BATTERIES

Figure 7.2 Recycling batteries protects the environment.

INSTRUCTIONS

EXERCISE 7 PARTS PER MILLION (PPM) AND OTHER TINY QUANTITIES

Last Name ____________________ **First Name** ____________________
Instructor ____________________ **Date** ____________________

Instructions. Use Table 7.1 to "translate" the US Environmental Protection Agency's (EPA) Maximum Contaminant Levels (MCLs) of the following drinking water pollutants, as shown for barium.

Table 7.2 Questions 1–10

Contaminant	Maximum Contaminant Level in ppm, etc.	Maximum Contaminant Level in g/L, etc.
Barium (Ba) Solved Example	2 ppm	2 mg/L
Copper (Cu)	1.3 ppm	____________
Lead (Pb)	0 ppb	____________
Nitrate (NO_3^-) and nitrite (NO_2^-) as nitrogen	10 ppm	____________
Chlorine (Cl_2)	4.0	____________
Fluoride (F^-)	4 ppm	____________
Arsenic (As)	10 ppb	____________
Beryllium (Be)	0.004 ppm	____________
Cadmium (Cd)	0.005 ppm	____________
Mercury (Hg)	0.002	____________
Selenium (Se)	0.05 ppm	____________

Questions.

11. What does this lab's information indicate about throwing used batteries in the trash?

12. If batteries are discarded in the trash, could the metals in them (cadmium, nickel, lead, etc.) migrate to drinking water?

13. What is the best way to dispose of used batteries?

14. Use Table 7.2 to rank in order of increasing toxicity (most dangerous last), the following: selenium, lead, copper, and mercury.

BIBLIOGRAPHY

www.epa.gov/ground-water-and-drinking-water/national-primary-drinking-water-regulations#inorganic

EXERCISE 8

Big and Small Numbers – Scientific Notation

One gram, the mass of a Sweet'n Low^{TM} packet, of carbon contains 50,000,000,000, 000,000,000,000 carbon atoms!

Figure 8.1 Even a one gram Sweet'N Low packet contains an amazing number of atoms.

DOI: 10.1201/9781003006565-10

INTRODUCTION

In Environmental Chemistry, we often deal with very large and very small numbers. In these cases, it can be easier to use scientific notation, such as 1×10^9 instead of standard notation which would be 1,000,000,000. Scientific notation also lessens the probability of an error. Wouldn't you rather write the number in the quote above as 5.0×10^{22}? Here are some solved examples for your practice.

INSTRUCTIONS

First study and practice the solved examples below. Next, answer the questions listed for you to work.

Example 1. Only 12 g (picture the contents of 12 Sweet'n Low™ packets such as those in Figure 8.1) of carbon, a mole of carbon, contains a staggering number of atoms: 602,000,000,000,000,000,000,000. This number can be converted to the more manageable scientific notation by simply moving the decimal point 23 places to the left and multiplying the 6.02 which results by 10^{23}. It is usually most convenient to "move" decimal points so that the final lead number is between 1 and 10, as in this example 6.02 is the first part of the number we obtained.

Example 2. To deal with very small numbers, such as the proton which is 0.00000000000000000000000167 g, we move the decimal point 24 places to the right, to get 1.67 and then because the decimal point was moved to the right that many places, we multiply the 1.67×10^{-24}, to yield 1.67×10^{-24} g.

Now, here are some more examples for you to work.

1. In a game of chess, it is said that the first ten opening moves can be played in at least 1.7×10^{26} ways. Express in standard notation.

2. The Pittsburgh Steelers National Football League team is valued at $3,000,000,000. Express this in scientific notation.

3. One gram of pitchblende (uranium oxide ore) contains less than 0.0000001 g of radium.
   a. Convert to scientific notation.

   b. Comment on the skill and diligence of Marie Sklodowska Curie, who isolated radium from pitchblende.

4. A porcupine has about 3.0×10^4 quills. Express this number in standard notation.

5. The electron mass is 9.11×10^{-28} g. Express in standard notation.

6. The best form of the fossil fuel coal is anthracite as it has the most energy content and burns without smoke. Its formation is said to require 440,000,000 years. No wonder it has been called a "non-renewable resource"! Express this length of time in scientific notation.

7. The size of the atom is sometimes measured in Angstrom Units (AU), which are 1×10^{-10} meter. The diameter of the helium (currently a scarce element) atom is about 0.64 AU. Express in meters in standard notation.

8. A 10K (10 kilometer) race involves 1,000,000 centimeter (cm). Express this number of centimeters in scientific notation.

9. Approximately 583,300,000,000 plastic bottles were sold worldwide in 2021. Express in scientific notation, and let's hope that most of these are recycled.

10. The speed of light is 3.0×10^8 m/s. Express in standard notation.

BIBLIOGRAPHY

Beard, James and Murphy, Ruth Ann, *Environmental Chemistry in Society*, Third Edition, Taylor and Francis Publishers, Boca Raton, FL, 2022, p. 138.

Tro, Nivaldo, *Chemistry: A Molecular Approach*, Fifth Edition, Pearson Publishing, Hoboken, NJ, 2020, p. 367.

history.aip.org/exhibits/curie/brief/06_quotes/quotes_07.html

https://www.chess.com>forum>view>general>fun-fact-first-ten-move-possibilities

https://www.forbes.com>teams>pittsburgh-steelers

www.livescience.com/56326-porcupine-facts.html

www.plasticsoupfoundation.org/en/2017/07/the-worlds-population-consumes-1-million-plastic-bottles-every-minute/

EXERCISE 9

Temperature and Heat – What's the Difference?

Venus is the hottest planet in the solar system, with a surface temperature of about 900°F.

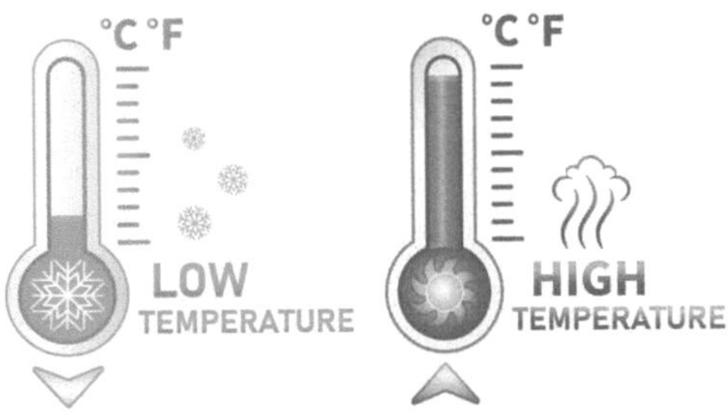

Figure 9.1 Degrees Celsius (°C) and degrees Fahrenheit (°F) are commonly used temperature scales.

INTRODUCTION

Temperature and **heat** are important topics! Surprisingly, the temperature of the planet Venus exceeds that of Mercury, even though Mercury is closer to the sun. This is attributed to the presence of carbon dioxide, a greenhouse gas, in the Venusian atmosphere. The CO_2 traps solar radiation, heating up the planet. With rising CO_2 levels and temperatures on our planet Earth, are we headed for destructive climate change? To understand this phenomenon, we need to understand temperature and heat. Temperature tells us how hot or cold something is. Heat energy will flow from objects at higher temperatures to objects at lower temperatures, if they are in contact. The temperature of the environment impacts climate change; our own temperature affects our health; and indoor temperatures may even precipitate arguments over thermostat settings! Some temperature units are shown in Table 9.1.

DOI: 10.1201/9781003006565-11

Table 9.1 A Comparison of Temperature Scales

Temperature Unit	Freezing Point of Water	Boiling Point of Water
Fahrenheit	32°F	212°F
Celsius	0°C	100°C
Kelvin	273 K	373 K

Here are some useful equations for changing temperature units:

$$°C = 5/9(°F - 32) \tag{9.1}$$

$$°F = 9/5(°C - 32) \tag{9.2}$$

$$K = °C + 273 \tag{9.3}$$

Similarly, **heat** impacts the Earth in many ways. Heat changes can be calculated with the following formula:

$$\text{Heat} = \text{Mass} \times \text{Temp. Change} \times \text{Specific Heat of the Substance} \tag{9.4}$$

Water has an amazingly high specific heat of 1.00 cal/g-°C, which is a survival mechanism for us. For comparison, the specific heat of the metal tin is only 0.239 cal/g-°C. Our bodies contain much water, and its high specific heat helps maintain our temperature at the proper level. This principle extends to our **hydrosphere** (all of the planet's water), in helping moderate temperatures. As an example, residents of coastal cities on average enjoy less harsh temperatures than those living further inland.

INSTRUCTIONS

Here are some solved examples of temperature conversion and heat calculation for you to practice. After you have done so, work the additional questions, remembering to show your method.

1. The surface temperature of the planet Venus, due in part to the greenhouse effect from atmospheric carbon dioxide, is about 900°F, even though Mercury, which is closer to the sun, is only about 800°F. Convert the 900°F temperature of the planet Venus to Celsius. Show method. [Hint: Use Equation (9.1) °C = 5/9 (°F – 32)]

Step 1. Write the equation. °C = 5/9 (°F – 32)
Step 2. Add the given number(s) to the equation. °C = 5/9 (**900** – 32)
Step 3. Simplify anything in parentheses. °C = 5/9 (**868**)
Step 4. Multiply and divide. °C = 5/9 (868) = 482°C

Comment: Although 482°C may *look* cooler than 900°F, it is the same – just in different units!

2. Convert the temperature of Venus, 482°C, to Kelvins. [Hint: Use Eqution (9.3) K = °C + 273]

Step 1. Write the equation. K = °C + 273
Step 2. Add the given number(s) to the equation. K = **482** + 273
Step 3. Add terms. Temperature = 755 K.

Comment: Kelvins are just that, Kelvins! They are basic units of nature and as such do not use the degree (°) notation.

3. Convert the Earth's coldest recorded temperature (about –90°C) in Vostok, Antarctica, to Fahrenheit. [Hint: Use Equation (9.2) °F = 9/5(°C) + 32]

Step 1. Write the equation. °F = 9/5 (°C) + 32)
Step 2. Add the given number(s) to the equation. °F = 9/5 (**–90**) + 32)
Step 3. Multiply and divide. °F = –162 + 32)
Step 4. Subtract. °F = –130°F (Brrr!)

4. Calculate the amount of heat required to warm a cup (229 g) of water from 25°C to 100°C. [Hint: Use Equation (9.4) Heat = Mass × Temp. Change × Specific Heat of the Substance]

Step 1. Write the equation. Heat = Mass × Temp. Change × Specific Heat of the Substance
Step 2. Add the given number(s) to the equation.

Heat = **229 g** × (**100°C** – **25°C**) × **1.00 calorie/g–deg**

Step 3. Simplify anything in parentheses.

$$\text{Heat} = 229\text{g} \times (\mathbf{75°C}) \times 1.00 \text{ calorie/g–deg}$$

Step 4. Multiply terms and cancel units where possible.

$$\text{Heat} = 229 \times (75) \times 1.00 \text{ calorie} = \mathbf{1718 \text{ calories}}$$

Heat = 229 g × 75°C × 1.00 calorie/g-deg = 17,175 calories
Note: These are not the same as nutritional calories, which are actually kilocalories (kcal), being a thousand times larger.

Now, try solving these. Remember to show your method.

1. Convert body temperature, 98.6°F, to Celsius.

2. Convert room temperature, 25°C, to Fahrenheit.

3. Convert –40°F, to Celsius. Show method and comment on your answer!

4. Convert 451°F, the temperature at which paper starts to burn, to Celsius.

5. Convert 649°C to Fahrenheit. The temperature of the average home fire exceeds this.

6. Compare your answers to Questions 4 and 5, and comment on the need for a fireproof safe in the home if important documents are stored there.

7. Convert 350°F (a common home oven baking temperature) to Celsius.

8. Convert 1000°C, a common hazardous waste incinerator temperature, to Fahrenheit.

9. The temperature in Death Valley in California can reach 56.7°C. Convert to Fahrenheit.

10. Calculate the quantity of heat required to raise the temperature of 1000 g of tomato soup from 20°C to 100°C. Assume the specific heat of tomato soup = 1.00 calorie/(gram °C) which is the value for water.

BIBLIOGRAPHY

Beard, James and Murphy, Ruth Ann, *Environmental Chemistry in Society*, Third Edition, Taylor and Francis Publishing, Boca Raton, FL, 2022, p. 351.

https://www.nps.gov>learn>nature>weather-and-climate

sf-fire.org/home-fire-facts

solarsystem.nasa.gov/planets/venus/overview/

EXERCISE **10**

Some Elements of Matcha Mole's Adventure in Environmental Chemistry Land

Using official symbols for chemical elements instead of their names can contribute to the efficiency of the environmental chemist.

Figure 10.1 A fresh spring meadow pictures an idyllic, well-managed environment with responsible use of chemicals.

DOI: 10.1201/9781003006565-12

INTRODUCTION

To properly care for the environment and to sustain flourishing outdoor areas such as shown in Figure 10.1 for agriculture, recreation, and more, chemicals must be used and managed by the environmental chemist. This task is made easier with the use of a systematic means of referring to chemical species. Today's environmental chemist can be more efficient by referring to chemical elements with their official symbols instead of writing their full names in words. As examples, the radioactive element darmstadtium is merely Ds, and rutherfordium, another radioactive metal, is Rf. And the simple (?) element carbon, is C. [Some of us do not consider carbon very simple, with concern about our "Carbon Footprint," which refers to the amount of greenhouse gases we generate.] These symbols for the elements may be one letter which is capitalized, such as C for carbon as previously mentioned, or two letters such as Rf for rutherfordium. Note that for two-letter symbols, only the first letter is capitalized. As chemistry is definitely an international endeavor, some symbols are based on the element's Latin name, such as K for potassium which has the Latin name kalium. In Exercise 11 we see another advantage of the use of these symbols; they streamline writing balanced chemical equations. This exercise provides practice in converting chemical symbols for the elements to their actual names.

INSTRUCTIONS

Use your lecture textbook to "decode" the following story by identifying the symbols of the elements, e.g., Sn would be tin.

Some Elements of Matcha Mole's Adventure in Environmental Chemistry Land

Last Name____________________ **First Name** ____________________
Instructor____________________ **Date** ____________________

Before driving to Environmental Chemistry class, I packed my new Ti ______________-graphite tennis racket for later, and took out my cell phone with a Li ____________________ battery to check messages. Then I noticed the fuel gauge was reading almost empty, so I stopped off at the gasoline station for some fuel with C______________ and H _______________. I paid with a credit card, not with Au______ or Ag________. Seeing some air pollution up the road ahead, I took a deep breath of O_2 _________ and closed the air vents on my truck. The sun with UV light was beaming through the sunroof, so I was glad I had put on some sunblock with Zn__________ before I left the house. Then there was this big traffic jam ahead and I hadn't had breakfast, so I grabbed some pretzels with Na_____________ to munch on while I waited, and waited! Discovered a sore tooth while eating, and decided to use more toothpaste with F______________ in the future. There was a big sale at the car dealer's lot across the street, with lots of bright red He__________________ filled balloons waving in the breeze. Everyone was still waiting in this traffic jam, so I pulled out a peanut butter sandwich I'd wrapped in Al___________________foil and began eating it. Finally, I got moving again, only to stop due to a flat tire; it had a nail made of Fe __________stuck in it! Trying to lift the spare, I found it heavy as Pb____________________, even though my bones are strong from all the Ca ______________________ and Mg____________________ I consume, but a police officer in a Co_________________ blue uniform happened to come by and changed the tire for me! I offered her some cold H_2O which I said was quite safe because it had been purified with Cl_2 _________________________.

Then someone ran into my truck's Cr ___________________ bumper and didn't even stop. I am glad it is not made of Cs ___________________ which is the softest metal and catches fire if wet.

Next, I noticed a Ne____________________ sign advertising free vehicle emissions checks, so I drove right in. Who wants to pollute?

They said my green truck passed, thanks to the catalytic converter with Pt ________________________accelerant.

When I got to Environmental Chemistry class, the instructor was already taking up lab reports, and in my haste to turn mine in, I got a paper cut, so I pulled out a little bottle of I_2 __________________________________ and dabbed it on the cut.

After class, I need to drop off some old light-emitting diode (LED) light bulbs at the recycling center – they may contain Cu _____________________, Ni ________________and lead but are much more efficient than the incandescent versions with W ________________________. I am also recycling some compact fluorescent light (CFL) bulbs as they contain Hg _____________________ and should never go in the trash!!

That's all for now!

Your Friend in Environmental Chemistry,

Matcha Mole

Figure 10.2 Matcha Mole, Your Environmental Chemistry friend. (Courtesy of Gabrielle Murphy Goldstein.)

EXERCISE 11

Environmental Chemical Reactions – Balancing Equations

*We are becoming more responsible for our "**Carbon Footprint.**"*

Figure 11.1 Our Carbon Footprint.

INTRODUCTION

We live in an era where *quantities* seem to be of increasing importance. Industries as well as individuals are becoming more responsible for, as shown in Figure 11.1, their "Carbon Footprint," i.e., the *quantities* of carbon dioxide and other greenhouse gases which they release to the environment due to their daily activities. Chemical industries also need to keep track of the amounts of chemicals used and produced to monitor profits and losses. As individuals, we

DOI: 10.1201/9781003006565-13

are also responsible for our care of the environment. Central to dealing with chemical quantities and their reactions is the process of balancing chemical equations. Types of reactions include **neutralization**, **oxidation-reduction**, and **precipitation**. Here are some examples. Neutralization reactions involve the transfer of hydrogen ions (H^+) and are exemplified by industries which must adjust the pH (acidity level) of their emissions to safe levels, by adding base to acid or acid to base. Oxidation-reduction reactions occur when electrons are exchanged between reactants, such as corrosion of metals and the resulting failure of structures they support. Precipitation reactions occur when hard water and soap combine to form a "bathtub ring," or two soluble substances in clear solutions react to form something insoluble as shown in Figure 11.2. Here are some equations for your practice in balancing.

Figure 11.2 Precipitation reaction of potassium chromate and lead nitrate to form gold lead chromate precipitate with potassium and nitrate ions remaining in solution.

INSTRUCTIONS

Balance the equations on the report sheet so that the numbers of each type of atom are the same on each side of the reaction arrow. Practice the examples below to understand the method; then complete the other equations.

Example 1. Sulfur dioxide, which is responsible for acid rain and other environmental problems, forms when sulfur-containing coal is burned, by the following reaction. Note that the equation is balanced without our adding anything; it already has exactly one sulfur atom on each side of the reaction arrow and two oxygen atoms. The engineering adage, "If it ain't broke don't fix it," holds true here, so the equation is balanced as given.

$$S + O_2 \rightarrow SO_2$$

Example 2. Carbon monoxide, which is formed when fuels are burned with limited oxygen, is a toxic gas. It can be converted to the much safer compound carbon dioxide by the following reaction. Here we begin with one carbon atom on each side, but there are three oxygen atoms on the left and only two on the right.

$$CO + O_2 \quad \rightarrow \quad CO_2$$

We can add a two in front of the CO to yield the following:

$$2CO + O_2 \quad \rightarrow \quad CO_2$$

Now, there are four oxygens on the left, two oxygens on the right, two carbons on the left, and one carbon on the right. By adding a two in front of the CO_2, as shown below, we can balance the carbons as well as the oxygens. There are now two carbons on each side and four oxygens on each side. Remember that a coefficient in front of a formula applies to everything in the formula! If the coefficient is understood to be unity (one), just do not write a coefficient as shown in front of O_2 below.

$$2CO + O_2 \quad \rightarrow \quad 2CO_2$$

Now, it's your turn!

EXERCISE 11 ENVIRONMENTAL CHEMICAL REACTIONS - BALANCING EQUATIONS

Last Name ____________________ **First Name**________________________
Instructor ______________________**Date** ___________________________

1. Burning coal for electricity converts the carbon of the coal to carbon dioxide as follows.
__________C + __________O_2 → ______________CO_2

2-5.
Hydrochloric acid, aka muriatic acid, is found in our stomach's gastric juice, and can be used for household cleaning, opening drain clogs, and many other purposes. Its neutralization by sodium hydroxide is shown below. Similarly, sulfuric acid, aka battery acid, and a component of industrial smog, can be neutralized by potassium hydroxide. Additional neutralization reactions are provided for your practice.

2. ______HCl + __________NaOH → ___NaCl + ___H_2O

3. ______H_2SO_4 + ___ KOH → ______K_2SO_4 + ______H_2O

4. ______HCl + ______$Ba(OH)_2$ → ______$BaCl_2$ + _____ H_2O

5. ______H_3PO_4 + ______$Ca(OH)_2$ → ______$Ca_3(PO_4)_2$ + ______H_2O

6-7.
An acidic environment, such as acid rain or even acid fog, can damage important metals, converting them to salts and destroying their structural integrity. Think corroded metal sculptures, failing building supports (rebars), etc. These are oxidation-reduction reactions, which involve electron transfer between reacting species. Here are examples of such reactions.

6. ________Zn + ______ HCl → ______$ZnCl_2$ + ______H_2

7. ________Al + _____ H_2SO_4 → ________$Al_2(SO_4)_3$

8. Oxidation-reduction reactions also occur when a more active metal, such as potassium, replaces a less active metal such as zinc. This principle is employed in the manufacture of **galvanized** steel; **steel** (iron plus carbon) is coated with zinc so the zinc will be oxidized, protecting the iron.

______K + _____$Zn(OH)_2$ → _____ KOH + _____ Zn

9. The famous Haber synthesis produces valuable ammonia from hydrogen and atmospheric nitrogen by the following reaction which is also an oxidation-reduction process. Ammonia can then be converted to fertilizer and explosives.

______N_2 + _____H_2 → ______NH_3

10. The interesting compound potassium perchlorate is used to inflate automobile airbags. It decomposes by an oxidation-reduction reaction as follows, producing oxygen gas.

    ________$KClO_4$ → ______ KCl + ___O_2

11. The highly insoluble substance calcium carbonate, or limestone rock, is part of the **carbon cycle** as it contains carbon. In this precipitation reaction, the subscripts (aq) for aqueous or dissolved and (s) for solid or precipitate are added to this equation to emphasize the phenomenon.

    _____$Ca(NO_3)_2$(aq) + _____Na_2CO_3 (aq) → _____$NaNO_3$(aq) + _____$CaCO_3$(s)

12. Carbonates such as limestone rock and even pearls can be dissolved by acid as shown in this equation. Never clean your pearls with vinegar!

    ___HCl + _____ $CaCO_3$ → _____$CaCl_2$ + _____CO_2

EXERCISE 11 ENVIRONMENTAL CHEMICAL REACTIONS – BALANCING EQUATIONS

Post-Lab

Last Name ________________________ **First Name**________________

Instructor ________________________ **Date** ____________________

Instructions. For multiple-choice questions, select the *best* answer.

1. Name the three types of chemical reactions discussed in this exercise.

2. Steel is made of
   a. iron
   b. iron and cesium
   c. iron and carbon
   d. zinc and iron

3. Galvanized steel is made of
   a. iron, carbon, and zinc
   b. iron and copper
   c. iron and carbon
   d. iron and nickel

4. The Haber process
   a. produces dolomite from calcium and magnesium
   b. produces steel from carbon and titanium
   c. produces ammonia from biomass
   d. produces ammonia from nitrogen and hydrogen

5. The advantage of galvanized steel over regular steel is its resistance to
   a. precipitation
   b. neutralization
   c. polymerization
   d. corrosion

6. Sulfuric acid, H_2SO_4, is also known as
   a. battery acid
   b. muriatic acid
   c. carbolic acid
   d. none of these

7. Hydrochloric acid, HCl, is also known as
   a. battery acid
   b. muriatic acid
   c. chloric acid
   d. none of these

8. Limestone rock is a form of compressed
   a. calcium bicarbonate
   b. calcium chloride
   c. calcium carbonate
   d. magnesium carbonate

UNIT B

Adventures in the Environmental Chemistry Lab

EXERCISE 12

Measurement – A Foundation of Environmental Chemistry

Experimenters are the shock troops of science.

– Max Planck, German scientist and Nobel Prize winner

Figure 12.1 Glassware used to measure volume in lab.

DOI: 10.1201/9781003006565-15

INTRODUCTION

Environmental Chemistry studies depend on measurements! We need to know the planet's temperature, as we deal with climate change. The concentration of salt in the ocean is important – as it can positively or negatively impact marine life. And many of us are interested in how many inches of rain our lawns receive.

The concentration of possible pollutants in our drinking water, our air, and our food are not just questions – they are vital for our health.

This lab introduces some ways to measure volume and temperature with the metric system, which is used and valued by scientists worldwide. Its basic units are the meter (m) for length, the kilogram (kg) for mass, and the liter (l) for volume. Sometimes the meter is inconveniently large, and centimeters (cm) which are hundredths of a meter are used. Instead of liters, the environmental chemist may opt for milliliters (mL), especially if measuring tiny quantities. As pointed out in Exercise 5 Exploring the Metric System, the United States is one of only three countries in the world which have not officially adopted the metric system, although we do use measurements such as 2 liters (2 L) for soft drink bottles, 10 kilometers (km) for races, and milligrams (mg) for products such as Vitamin C supplements. It is helpful to remember that a liter is just a bit larger than a quart (1.00 L = 1.06 qt), and a meter is just a bit longer than a yard (1.00 m = 39.37 in.) Likewise, as an approximation, 1.00 kilometer (km) = 5/8 of a mile. Our lab work usually involves temperature measurements with the Celsius Scale, which has 100 degrees between the freezing point and the boiling point of water. The Fahrenheit scale has 180 degrees between these two temperatures.

In this lab, we measure volumes (Figure 12.1), masses, and temperatures. Volumes are measured with graduated cylinders, and approximated with graduated beakers, which are less accurate. **Mass** is often referred to as **weight**, although mass is the actual amount of material present, whereas weight depends on the gravitational field where the measurement is made. Mass is determined with the use of a **balance**, which is the preferred term instead of scales which would indicate a less accurate instrument.

Our measurements in this lab also involve salt (sodium chloride), which has many uses in our society. While we tend to regard salt as common, in times of the Roman empire, salt was expensive. We get our word "salary" from the Latin word "salarium" which was the payment soldiers received to purchase this valuable commodity. Salt can enhance the flavor of foods as well as help preserve them. Salt is also used on highways to prevent the formation of ice, as well as in the freezing mixture to make homemade ice cream. Both processes depend on the lowering of the freezing point of water by the presence of sodium chloride, an **electrolyte**. Electrolytes conduct electrical current when melted or dissolved in water.

SAFETY

Put on your PPE. Although the items used in some labs may appear innocuous, proper safety considerations dictate that PPE be worn in a science lab.

SUPPLIES FOR 24 STUDENTS OR TEAMS

Six balances centigram or better, 24 rulers marked in centimeters (cm), 24 250 mL graduated beakers, 24 100 mL graduated cylinders, weighing boats, box of table salt, 24 non-mercury Celsius thermometers such as –10°C to 110°C, 24 glass stirring rods, vial of red litmus paper, vial of blue litmus paper, crushed ice.

When using the balance, turn on the balance and gently place the item to be "weighed," in the middle of the balance pan. Allow time for the balance to stabilize. Read the display and record the value. If anything is spilled on a balance, your instructor should be notified promptly to avoid damage to this sensitive and often expensive device. Each balance has its limit as to the maximum weight it can handle, so to avoid damage, never place anything on the balance which is not specifically mentioned in the experiment. Never place chemicals directly on the balance as they could cause corrosion. Never place a hot object on a balance.

Volumes can be measured with graduated cylinders more accurately than with graduated beakers. The latter have markings for convenience but are less exact, as pointed out above.

When measuring temperature with a thermometer, keep the thermometer vertical and do not allow it to touch the sides or bottom of the container. Thermometers that are not kept vertical may develop a separation of the liquid in their column.

EXERCISE 12 – MEASUREMENT – A FOUNDATION OF ENVIRONMENTAL CHEMISTRY

Pre-Lab Questions

Last Name ____________________ **First Name** _________________
Instructor _____________________ **Date** ______________________

1. True or False? The United States official measurements are based on the metric system.

2. When liters are too large a unit, ________________ are commonly used.
3. Which is true of a liter?
   a. Equal in size to a quart
   b. Equal in size to a pint
   c. Slightly larger than a quart
   d. Slightly smaller than a quart
4. The precise instrument used to determine mass in a lab is properly called
   a. scales
   b. a balance
5. Rank the following by length, longest first
   a. kilometer > meter > mile > centimeter
   b. kilometer > mile > centimeter > meter
   c. mile > kilometer > meter > centimeter
   d. kilometer > meter > mile > centimeter

EXERCISE 12 MEASUREMENT – A FOUNDATION OF ENVIRONMENTAL CHEMISTRY

Data Page

Last Name ________________ **First Name** ______________
Instructor ________________ **Date** ________________

1. *Measurement of a lab drawer.* Measure the length, width, and depth of your lab drawer in cm and record the results below. [For length, you may use the partial length which allows access to contents without removing the drawer.] Convert to meters by dividing each result by 100, and record.
   Length of lab drawer ________________ cm ________________ m
   Width of lab drawer ________________ cm ________________ m
   Depth of lab drawer ________________ cm ________________ m

2. *Calculation of area of lab drawer.* Multiply the length and width from Question 1 and record below – the units will be in square centimeters (cm^2). Convert to square meters (m^2) by dividing by 100^2 which is 10,000 and record.

   Area of lab drawer ________________ cm^2 ________________ m^2

3. *Calculation of volume of lab drawer.* Multiply the area in cm^2 from Question 2 by the depth from Part 1 and record below – the units will be in cubic centimeters (cm^3). Convert to cubic meters (m^3) by dividing by 100^3 which is 1,000,000 and record.

   Volume of lab drawer ________________ cm^2 ________________ m^3

4. *Measurement of volume of a graduated beaker.* Fill a 250 mL graduated beaker with tap water to the 50 mL mark. Transfer to a 100 mL graduated cylinder and read and record the liquid level at the bottom of the meniscus, holding the cylinder at eye level to avoid parallax error.

   Record the volume. ________________ mL

   Account for any differences in the measurements.

   ____

5. *Measurement of mass.* Place an empty weighing boat on the balance and record the mass. Then slowly add sodium chloride until the balance registers a mass of about 5.00 g more than that of the weighing boat. Complete the table below. Look at the salt – that is the amount in one can of some types of soup. Save the salt for another part of this experiment.
   Mass of weighing boat with salt ________________ g

   Mass of empty weighing boat ________________ g

   Mass of salt ________________ g

Weigh the following objects and record your results.

Mass of one strip of red litmus paper (turns blue in base) _______________g

Mass of one strip of blue litmus paper (turns red in acid) ________________g

Mass of another strip of blue litmus paper ___________g

Are the above three values the same? _________ If not, how would you account for any differences? ____

6. *Measurement of temperature*. Fill the 250 mL beaker about 2/3 full with tap water, place the thermometer in the liquid, allow a couple of minutes for it to stabilize, and record the temperature. ________°C

Empty the beaker, fill it about 2/3 full with crushed ice, add tap water till the ice is just covered, and measure and record the temperature as you did for tap water _________°C

Add the salt you weighed out in Part 5 to the beaker, stir well, and measure and record the temperature. _________°C

How do the last two temperatures compare?

____

How would you explain any difference in the temperatures of the last two samples?

____

____

Rinse out your glassware including the weighing boat and dry it before returning it to your lab drawer. Dispose of the salt water and litmus paper as directed by your instructor.

BIBLIOGRAPHY

Erlich, Eugene, *Amo, Amas, Amat and More: How to Use Latin to Your Own Advantage and to the Astonishment of Others*, Harper & Row, New York, 1985, p. 92.

EXERCISE 13

Why Do Icebergs Float? A Study of Density

Mercury, a neurotoxin which is now banned from many uses in our society, has a density of 13.6 g/mL, almost 14 times that of water.

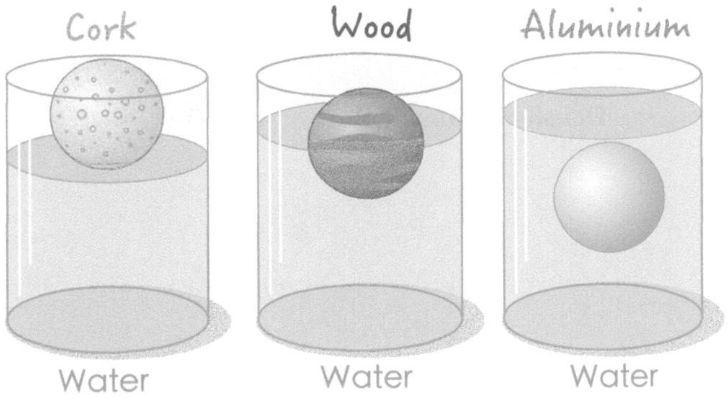

Figure 13.1 The density of a substance affects what level it will reach in water. [Aluminium is the alternative spelling for the common metal aluminum.]

INTRODUCTION

The properties of substances are dependent on their densities. **High-density polyethylene** (HDPE) is used for containers such as plastic bottles, toys, and pipes. In contrast, **low-density polyethylene** (LDPE) is used in large quantities for plastic bags, bubble wrap, etc. Densities account for the fact that if we go fishing, a cork (density = 0.35 g/mL) will float in the water, whereas a sinker will, as its name implies, sink. Objects float in water if their density is less than 1.00 g/mL which is the density of water, and objects sink in water if their density exceeds the density of water. Of the three physical states of matter – solid, liquid, and gas – gases have the lowest densities,

DOI: 10.1201/9781003006565-16

which are expressed in grams per liter, while the densities of solids and liquids such as shown in Figure 13.1 are expressed in grams per milliliter. Different liquids may have quite different densities, enabling them to form layers, as when oil spills occur in the oceans and the oil floats on top of the water, and as shown in Figure 13.2

Lakes freeze with the ice at the top, because ice is less dense than liquid water – this also explains why ice cubes *float* rather than *sink* in our beverages. Water is very unusual in this respect, as the solid states of most substances are denser than their liquid state. Oil spills at sea are found on top of the water because oil is less dense than water. So, what is density, anyway? Density, as explained in Exercise 6 Mastering the Metric System, can be defined as the mass of a given amount of a substance, divided by its volume. Density is called an **intrinsic property**, because, unlike mass and volume which are **extrinsic properties**, its value does not change with the amount of the substance being studied. To determine densities, we must measure both mass and volume. Mass determinations require careful use of a laboratory balance. Volume measurements require graduated glassware with the lowest part of the meniscus (dip in the liquid surface) read while it is at eye level. Let's explore this topic!

$$\textbf{Density = Mass/Volume} \quad \textbf{or} \quad D = \frac{m}{v}$$

Figure 13.2 A density column showing that liquids form layers with the denser substances lower in the container.

EXERCISE 13 WHY DO ICEBERGS FLOAT? A STUDY OF DENSITY

Pre-Lab Questions

Last Name ____________________ **First Name** ________________
Instructor _____________________ **Date** ______________________

1. Define density.

2. Select the common units of density for solids and liquids.
   a. grams
   b. milliliters
   c. grams per milliliter
   d. milliliters per gram

3. When determining the level of a liquid in a graduated cylinder, the liquid level should be
   a. above eye level
   b. at eye level
   c. below eye level
   d. none of these

4. When determining the level of a liquid in a graduated cylinder, it is important to read the
   a. highest part of the meniscus
   b. lowest part of the meniscus

5. Arrange the following in order of increasing density, with the least dense form first.
   a. gaseous water < solid water (ice) < liquid water
   b. gaseous water < liquid < solid water (ice)
   c. liquid water < solid water (ice) < gaseous water
   d. liquid water < gaseous < solid water (ice)

INSTRUCTIONS

Supplies for 24 students or teams. 6 balances, centigram or better, 24 100 mL graduated cylinders, 24 10 mL graduated cylinders, 24 weighing boats (may be rinsed, allowed to dry, and reused), unknowns such as diet and regular sodas, salt water, sugar water, colored water, vinegar, copper shot, aluminum shot, tin shot, iron nails or tacks, glass beads. Solid unknowns can be allowed to dry on paper toweling and reused in future experiments.

Safety: Put on your PPE.

Density of a liquid. Get an unknown from your instructor and record its number on the data sheet.

Weigh a dry, empty 10 mL graduated cylinder and record its mass on line 2 of Table 13.1.

Next, add some of your unknown up to the 10 mL line and weigh again, recording the mass on line 1 of Table 13.1. Note that the liquid level will assume a curved shape – this is called the **meniscus**. Place the graduated cylinder at the same level as your eye to avoid parallax errors, and adjust the liquid level with a Beral pipet so that the lowest portion of the meniscus just touches the 10 mL mark.

Wash and dry the cylinder and repeat the study for trial 2.

Density of a solid. Get an unknown from your instructor and record its number on the data sheet.

Weigh the weighing boat and record its mass on line 2 of Table 13.2.

Place about 15 mL (looks like a tablespoonful in a measuring tablespoon) of your unknown in the weighing boat and weigh this, recording the mass on line 1 of Table 13.2.

Subtract to get the mass of your unknown, recording it on line 3 of Table 13.2.

Add about 40 mL of tap water to your 100 mL graduated cylinder, read the level of the liquid, and record on line 5 of Table 13.2.

Add your solid unknown carefully to the 100 mL graduate cylinder with the water in it, being careful not to spill any of the solid or splash any of the water out.

Read the new level of the liquid in the cylinder and record on line 4 of Table 13.2.

Subtract line 4 from line 5, and enter the answer on line 6 of Table 13.2. This is the volume of the solid you added.

Divide line 3 by line 6 and write the answer on line 7 of Table 13.2. This is the density of the solid.

Rinse and dry your graduated cylinder. Get a fresh sample of the same unknown and repeat the experiment entering your results for trial 2 in Table 13.2.

When finished, rinse and dry your glassware and return it to the designated area. Place all of your samples in the areas designated by your instructor.

EXERCISE 13 WHY DO ICEBERGS FLOAT? A STUDY OF DENSITY

Data

Last Name ______________________ First Name ______________________
Instructor ______________________ Date ______________________

Table 13.1 **Determination of the Density of an Unknown Liquid**

Liquid Unknown Number ________ Density of Unknown Liquid	Trial 1	Trial 2
1. Mass of graduated cylinder with 10.0 mL of unknown liquid (g)		
2. Mass of dry, empty 10 mL graduated cylinder (g)		
3. Mass of unknown liquid (line 1 minus line 2 in g)		
4. Density of unknown liquid in g/mL. Divide g of unknown liquid in line 3 by 10.0 mL		

Table 13.2 **Determination of the Density of an Unknown Solid**

Solid Unknown Number ________ Density of Solid	Trial 1	Trial 2
1. Mass of solid in weighing boat (g)		
2. Mass of empty weighing boat (g)		
3. Mass of solid (line 2 minus line 1) (g)		
4. Final volume of water in graduated cylinder after solid was added (mL)		
5. Initial volume of water in graduated cylinder before solid was added (mL)		
6. Volume of solid (line 4 minus line 5) (mL)		
7. Density of solid (line 3 divided by line 6) (g/mL)		

EXERCISE 13 WHY DO ICEBERGS FLOAT? A STUDY OF DENSITY

Post-Lab Questions

Last Name ____________________ **First Name** ____________________
Instructor ____________________ **Date** ____________________

1. How closely did your values for the density of your liquid unknown agree? Explain.

2. How closely did your two trials for the density of the solid agree? Explain.

3. Could the method used in today's experiment be used to determine the density of a cork? Explain.

4. Assuming someone splashes about 5 mL of water out of the cylinder when adding the solid unknown, would this cause an experimental error? If so, would the measured density be too large or too small?

5. List two ways our environment would be changed if the density of ice were greater than that of liquid water
   a.

   b.

EXERCISE 14

Metals and Alloys – What's the Difference?

Group 11 of the Periodic Table includes "The Coinage Metals," copper (Cu), silver (Ag), and gold (Au).

Figure 14.1 Pennies contain copper, one of the three coinage metals, the other two being the precious metals silver and gold.

INTRODUCTION

Metals in our environment are used for many purposes in addition to minting coins (Figure 14.1), including the manufacture of vehicles, and support for structures like bridges and buildings. Some of these metals are more reactive than others, and thus vulnerable to environmental pollutants in the form of acid rain and more. In this lab, we study metals, alloys, and their reactions.

DOI: 10.1201/9781003006565-17

You may have noticed that most of the elements on the periodic table are **metals**. In the periodic table below in Figure 14.2, the metals are shown to the left of the diagonal formed by the elements B, Si, Ge, As, Sb, Te, and At. Metals also comprise the lower two rows of the periodic table, which begin with La and Ac and are called the Lanthanide Series and the Actinide Series, respectively. So, there are lots more metals than there are non-metals.

Metals reflect light, are **ductile** (can be drawn into a wire), and are good conductors of heat and electricity. Except for a few like copper and gold, they are all silvery or gray in color. Although they have many similar properties, each metallic element is unique. Some react vigorously with water while others do not. Some are soft and easily pounded into sheets (**malleable**) while others are hard and brittle. No two elements have exactly the same properties.

Mixtures of metal elements in specific proportions are called **alloys**. Alloys are solid solutions which have properties that are often very different from their elements and are thus very useful. The popular jewelry metal **rose gold** is an alloy of 75% gold and 25% copper. An alloy of lead and tin, called **solder**, has a melting point lower than either element by itself. Alloys of mercury (the *liquid* metal) with silver are called **amalgams** and used for dental fillings, although this use of mercury is a concern to some. Various properties can be designed into an alloy by changing the amounts of the elements making up the mixture.

Part I. Phase Changes of Copper Wire and Nitinol Alloy

There are a variety of phases: solids, liquids, and gases are the most common phases; the liquid crystalline state possessed by some materials is used in a variety of display devices (**liquid crystalline display** or **LCD**).

Many solids undergo phase changes that result in a change in structure at the atomic level but leave the material in a solid form. A particularly remarkable example is a compound called **memory metal** or **shape memory alloy** (SMA). The SMA example that will be demonstrated for you is an alloy of Ni (nickel) and Ti (titanium) that is sometimes called **Nitinol** (NiTiNaval-Ordinance Laboratory, after its elements and the lab where it was discovered).

Some applications of SMAs already exist: shower fixtures that shut off if the water becomes too hot, protecting a person from scalding water; frames for glasses that regain their shape after being bent; and tight seals between an SMA and another piece of material. A potentially huge market would be automobile parts that could regain their shape after an accident simply by warming. Maybe you can think of some new use for these materials.

At very high temperatures, Nitinol can be molded into a desired shape. At room temperature, the solid can be bent into a second shape. Modest heating to about 50 or 60°C causes a phase change in which the metal transforms itself back into its

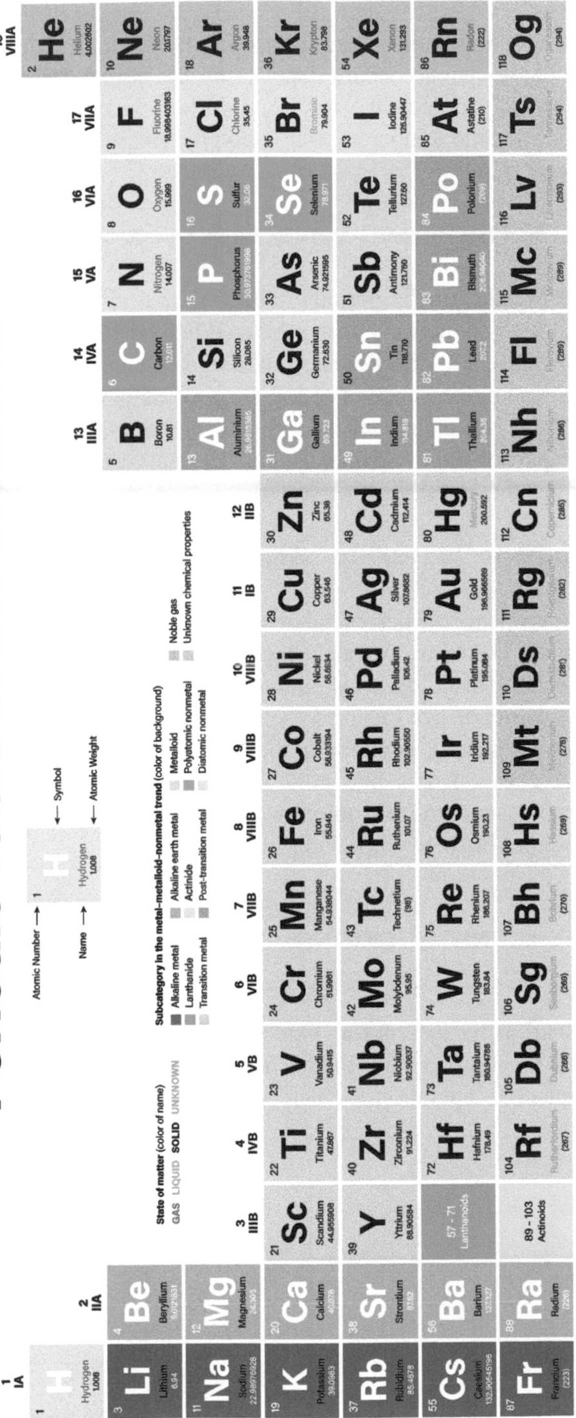

Figure 14.2 The periodic table of the elements is composed mainly of metals. Non-metals are on the far right, ending approximately with those on the diagonal line from boron (B) through oganesson (Og).

original shape, i.e., its "memory" is activated. Structurally, the atoms of the solid only have to shift positions slightly in order to regain its original shape. The temperature at which the phase change occurs depends on the composition of the metal: replacement of Ni with Co (cobalt), for example, to make alloys of the composition $Ni_xCo_{1-x}Ti$ permits the temperature at which the phase change occurs to be "tuned" to the value of x.

Part II. Reaction of a 1982 (or Newer) Penny with Hydrochloric Acid

Among "The Coinage Metals" of the periodic table, silver and gold coins are rare these days, so we will study pennies. Those produced before 1982 were made mostly from copper (95% Cu and 5% Zn). Pennies made since then are made with a zinc core and only a thin coating of copper (97.5% Zn and 2.5% Cu), as a cost-savings measure. One of these two metals reacts with hydrochloric acid (HCl), but the other does not. In this part of the experiment, you will determine which metal reacts and which is left after the reaction.

Part III. Changing Copper to "Silver" to "Gold"

In this part of the experiment, copper metal will be plated with zinc metal, giving it a silvery appearance. Then the zinc-coated copper will be heated so that an alloy of copper and zinc forms. This alloy is called **brass** and has the appearance of gold. In this part of the experiment, we must use "old" pennies (before 1983) because we need the coin to be mostly copper for the brass to form. The ancient alchemists tried to change cheap metals into gold. We can change pennies into something that *looks like* gold! And you just might want to save your pennies - the US Mint announced cessationn of their production in 2023.

INSTRUCTIONS

Supplies for demonstration by instructor: One piece of Nitinol wire, one piece of copper wire (for comparison), tongs, hot air gun, safety shield, Bunsen burner to demonstrate operation for Part 3.

INSTRUCTIONS FOR PART I. PHASE CHANGES OF COPPER WIRE AND NITINOL ALLOY

Safety: Put on your PPE.

Your instructor will perform the following demonstration of some of the properties of this SMA. Watch the demonstration and answer the questions on your data page.

1. The demonstration will be done first on the copper wire, then on the Nitinol wire.
2. Take the copper wire and bend it to demonstrate its stiffness (or lack of same).
3. Fold the wire into a reasonably compact shape.
4. Holding the wire with tongs, heat the wire and observe any changes that take place.
5. After the wire has cooled, repeat steps 2–4 for the Nitinol wire.

INSTRUCTIONS FOR PART II. REACTION OF A 1982 (OR NEWER) PENNY WITH HYDROCHLORIC ACID

Supplies for 24 students or teams of students: Twenty-four 1982 (or later) pennies, twenty-four 2 fl. oz. (about 60 mL) plastic cups with lids, twenty-four triangular files, about 500 mL of dilute (6 M) hydrochloric acid (HCl).

Caution: Corrosive.

1. Use the triangular file to completely remove the copper from most of the edge of the penny. This will expose the silvery zinc. Leave a few spots where the copper is not removed.
2. Write your name(s) on the plastic cup using a pen or marker.
3. Put the penny in the cup and cover it with about 1 cm (1/2 inch) of hydrochloric acid (6 M HCl). **CAUTION: CORROSIVE. Avoid eye and skin contact**.
4. Record on your data page what is happening in the cup.
5. Place the cup in the location designated by your instructor. The cup will be allowed to sit until next week's lab to allow time for the reaction of the hydrochloric acid with one of the metals. (While many chemical reactions are rapid, some reactions, like this, which somewhat illustrates the effects of acid rain on metallic structures, are *slow*.)
6. (To be completed next week) Write on your data page the appearance of the penny.

____

____

INSTRUCTIONS FOR PART III. CHANGING COPPER TO "SILVER" TO "GOLD"

Supplies for 24 students or teams: 72 pennies (48 from 1982 or earlier), 24 Bunsen burners, 24 evaporating dishes, fume hoods, 24 tongs, 100 g mossy zinc metal, 600 mL sodium hydroxide solution (3 M NaOH) Caution: Corrosive, 24 pencil erasers, paper towels, one or more beakers with tapwater, receptacle for wet paper towels, 24 small zip-lock bags to function as coin holders.

1. Clean the old pennies (and the new one, if needed) with the eraser of a pencil. Be sure to get the penny thoroughly clean or the zinc will not attach well to the penny.
2. Set one penny (the new one) aside. This is the "copper" penny.
3. Place an evaporating dish on a hot plate in the hood. Add about 5 mL of zinc to the dish. Carefully pour sodium hydroxide solution over the zinc until it is covered. **CAUTION: Sodium** hydroxide is corrosive. Avoid eye and skin contact. All chemicals in the hood should be at least six inches from the sash. Using tongs, take the two old, cleaned pennies and place them into the evaporating dish in the hood. **The evaporating dish containing a hot mixture of zinc metal and sodium hydroxide solution is very corrosive, so be careful not to splash it as you put the pennies in the evaporating dish.** Carefully heat the evaporating dish until you can see steam.
4. Once the pennies are coated with zinc, remove them from the evaporating dish using the tongs. While still holding the pennies with tongs, wash them with cold water in a beaker in the hood next to the dish containing the zinc. Make sure pennies remain in beaker if transporting them away from the hood. Carefully blot the pennies dry with a paper towel; if you rub too hard, you will peel the zinc coating off the penny. **CAUTION: After blotting the pennies do not throw the paper towel in the trash basket. Instead, thoroughly wet the towel with water by immersing it in the designated receptacle in the waste hood. Your instructor will take care of disposing of these paper towels.**
5. Set one of the zinc-coated pennies aside. This is the "silver" penny.
6. Connect the Bunsen burner hose to the gas jet. Carefully turn on the gas and light the burner, adjusting the air and fuel intake to achieve a blue flame with a blue inner cone. Too little air will give a yellow flame. Pick up the other zinc-coated penny by the edges with the tongs and heat it gently in the Bunsen burner flame until the surface of the penny turns a uniform golden color. DO NOT OVERHEAT!! This is now a brass-coated penny.
7. Quickly drop the brass-coated penny into cold water to cool it before handling. Again, be careful drying the penny, to avoid removing the brass coating. This is the "gold" penny.
8. Put the three pennies (copper, "silver," and "gold," in the coin holder provided, add your name, and seal or staple it closed to protect your pennies. Submit to your instructor.
9. Answer the questions on your data page.

Instructor's initials for completed coins:____________________________

EXERCISE 14 METALS AND ALLOYS – WHAT'S THE DIFFERENCE?

Pre-Lab

Last Name ____________________ **First Name** ________________
Instructor ____________________ **Date** ____________________

1. Metals are good conductors of ________________ and ________________.

2. All metals are silvery in color except ______________ and ______________.

3. Define an alloy and give an example.

   ____

4. What are the most common phases materials possess?

   ____

5. What does SMA represent?

   ____

6. Name two common applications where SMA exists in our everyday lives.

   ____

7. What acid will be used in this experiment? Write its chemical formula.

   ____

8. What base will be used in this experiment? Write its chemical formula.

   ____

9. What precautions should be taken when handling both this acid and this base? (You may need to refer to your safety rules and guidelines for the laboratory!)

   ____

EXERCISE 14 METALS AND ALLOYS – WHAT'S THE DIFFERENCE?

Data

Last Name __________________ **First Name** _______________
Instructor ____________________ **Date** ____________________

Part I. Phase Changes of Copper Wire and Nitinol Alloy

1. Were the wires easily bent?
   a. Copper ____
   b. Nitinol ____

2. What happened when the wires were heated?
   a. Copper ____
   b. Nitinol ____

Part II. Reaction of a 1982 (or Newer) Penny with Hydrochloric Acid

3. Describe what you observed in the cup when the acid was added.

   ____

4. Did the copper react (dissolve)? _________ Did the zinc react (dissolve)?

   ____

5. Circle the reaction which you observed.
   a. $Cu(s) + 2H^+ (aq) \rightarrow H_2 (g) + Cu^{2+} (aq)$
   b. $Zn(s) + 2H^+ (aq) \rightarrow H_2 (g) + Zn^{2+} (aq)$

6. Describe the appearance of the penny after it was left a week in hydrochloric acid.

   ____

7. Which metal, copper or zinc, was dissolved by hydrochloric acid?

   ____

8. Based on your observations in this lab, what effects might acid rain have on certain metals in our society, weakening structures, damaging art, etc.?

   ____

Part III. Changing Copper to "Silver" to "Gold"

9. Describe what happened when the pennies were placed in the zinc and sodium hydroxide solution mixture.

 ____

10. Describe what happened when one of the pennies that had been treated with zinc and sodium hydroxide was heated.

 ____

EXERCISE 14 METALS AND ALLOYS – WHAT'S THE DIFFERENCE?

Post-Lab Questions

Last Name ____________________ **First Name** ________________
Instructor ____________________ **Date** ____________________

1. What is the difference between a pure metal and an alloy?
____

2. True or False: All metals react with hydrochloric acid (HCl) in the same manner.
____

3. What metal is used as the core for pennies made after 1983?
____

4. Why was the metal core changed?
____

5. What is the natural color of most metals?
____

6. The ancient alchemists tried to change cheap metals into
____

7. Copper, silver, and gold in Group 11 of the periodic table are known as the
____

8. Dental amalgams, used for fillings, contain silver and the liquid metal
____

9. Looking at the name of the iron alloy *alnico* and the symbols for metals, identify the other three metals in the alloy.
____

10. List six properties of metals discussed in this lab.
____
____

BIBLIOGRAPHY

https://www.newswire.com/news/what-will-happen-when-the-u-s-stops-making-pennies-in-2023-21409166

EXERCISE 15

Recycling Copper – Learning Reactions

Copper derives its name from the Latin word cyprium because in the past most copper came from the island of Cyprus.

Figure 15.1 Copper and its alloy bronze are produced in huge quantities.

INTRODUCTION

The supply of metals on the Earth is limited. Exercise 14 in this text explains why the copper concentration of pennies has been decreased dramatically. Copper is a valuable metal; some of its many uses are as wiring for electrical conduction and as corrosion-resistant pipe. It is used for roofing and, due to its antimicrobial properties, it is used in kitchens. Copper jewelry is popular as is rose gold, one of its alloys. It makes sense to recycle metals when possible. In this experiment, you get to "recycle" copper – by dissolving elemental copper wire and carrying out a series of five consecutive chemical reactions to recover the pure copper metal. It is a particularly interesting experiment because each new copper species that is produced can be easily identified visually by one or more of its characteristic physical properties. It also

DOI: 10.1201/9781003006565-18

provides a very good measure of your laboratory ability since you are attempting to recover all of the copper with which you began the experiment. [You cannot, of course, recover 100% of the copper since some is invariably lost in certain operations, such as decanting, precipitation, etc., but you can minimize such losses with good laboratory technique. It is also possible to obtain a yield greater than 100% if the sample contains water because it was not thoroughly dried, or if other impurities are present.] The single most important thing to keep in mind is that the loss of any copper-containing species in any step of the experiment will result in a lower yield of pure copper in the final step, so follow instructions carefully. The sequence of chemical changes is shown in Table 15.1 by names, chemical formulas, and appearances (what you should look for as you carry out this experiment). Being a transition metal, copper forms many colorful compounds, and is the basis for the beautiful turquoise jewelry of the American Southwest!

Table 15.1 The sequence of reactions for this lab is shown by words, chemical formulas, and the appearance of the chemicals

Copper	→	Copper nitrate	→	Copper hydroxide	→	Copper oxide	→	Copper sulfate	→	Copper
Cu(s)	→	$Cu(NO_3)_2(aq)$	→	$Cu(OH)_2(s)$	→	CuO(s)	→	$CuSO_4(aq)$	→	Cu(s)
Orange-red metal	→	Blue solution	→	Blue precipitate	→	Black preci-pitate	→	Blue solution	→	Orange-red metal

INSTRUCTIONS

Step 1. **Conversion of copper metal to copper (II) nitrate**

Weigh about 1 g of copper wire to 0.01 g and record the weight in the data section. Place the copper in a 250 mL beaker and add 5 mL of 16 M (concentrated) nitric acid (HNO_3). Be especially careful with acid and base solutions, protecting your eyes with goggles, and in case you get any on your skin, wash it off with plenty of water. Nitric acid, if spilled on the skin, can not only burn, but leave yellow spots which have to wear off with time. This is due to the formation of compounds called xanthoproteins ("yellow proteins").

Warm the solution gently on a hot plate in the hood until the metal is completely dissolved, the brown nitrogen dioxide gas NO_2 has been driven off, and the solution is deep blue in color. Obtaining a blue color in the solution is one of the most important steps to success in this experiment. A green color indicates that yellow-brown, nitrogen dioxide, NO_2, fumes are still dissolved in the blue copper (II) nitrate, $Cu(NO_3)_2$, solution and must be removed to prevent complications in later steps.

The chemical reaction carried out in this step in which the copper metal is put into solution is an **oxidation-reduction reaction** (electrons are transferred), which is as follows:

$$Cu + 4HNO_3 \rightarrow Cu(NO_3) + 2NO_2 + 2H_2O$$

Step 2. **Conversion of copper (II) nitrate to copper (II) hydroxide**

Cool the solution from Step 1 in an ice-water bath and slowly add about 10 mL of 6 M sodium hydroxide (NaOH) **CAUTION: CORROSIVE. Avoid contact with eyes and skin.** Stir the solution thoroughly and test it with red litmus paper, which will turn blue with base. If it is not basic, as indicated by a red-to-blue indicator paper color change, add more sodium hydroxide until it does turn red litmus paper blue. Remember to stir the solution thoroughly each time before testing. The pale blue solid that has formed is copper (II) hydroxide, $Cu(OH)_2$.

The chemical reaction carried out in this step in which copper (II) nitrate is reacted with sodium hydroxide to produce copper (II) hydroxide and sodium nitrate is a **metathetical reaction** (ions switch partners), which is as follows:

$$Cu(NO_3)_2 + 2NaOH \rightarrow \quad Cu(OH)_2 + 2NaNO_3$$

Step 3. **Conversion of copper (II) hydroxide to copper (II) oxide**

To the beaker containing $Cu(OH)_2$ from part II, add 100 mL of distilled water and boil gently on the hot plate while stirring constantly until the blue copper (II) hydroxide is converted to black copper(II) oxide, CuO, and water in a **dehydration reaction**. On line 6 of your data page, write the chemical reaction for this step in which copper (II) hydroxide reacts to form copper (II) oxide and water.

Step 4. **Conversion of copper(II) oxide to copper (II) sulfate**

Allow the copper (II) oxide precipitate from Step 3 to settle; then decant (carefully pour off) two-thirds of the liquid. Add 10 mL of dilute sulfuric acid (H_2SO_4) **CAUTION: CORROSIVE**, until the CuO precipitate has disappeared to yield a blue solution of copper (II) sulfate.

On line 7 of your data page, write the metathetical chemical reaction for this step in which copper (II) oxide reacts with sulfuric acid to produce copper (II) sulfate and water:

Step 5. **Conversion of copper (II) sulfate to copper metal (completion of cycle)**

Add about 2 g of zinc metal to the copper (II) sulfate solution from Step 4 and allow it to stand, stirring occasionally until the blue color has disappeared, or your laboratory instructor indicates that you should proceed to the next step in the experiment. If all copper (II) ions have been converted to copper atoms, the solution will now be colorless. On line 8 of your data page, write the oxidation-reduction

Safety. Put on your PPE.

Supplies for 24 students or teams of students: 24 g copper wire, 6 balances to 0.01 g or better, 24 250 mL beakers, 24 50 mL beakers, pencils, 200 mL concentrated nitric acid CAUTION: CORROSIVE, 12–24 hot plates, 2–3 fume hoods, 400 mL 6 M sodium hydroxide CAUTION: CORROSIVE, red litmus paper, 24 glass stirring rods, ice water bath, deionized or distilled water, 400 mL dilute sulfuric acid CAUTION: CORROSIVE, 60 g zinc metal, 400 mL dilute (6M) hydrochloric acid, drying oven at 110 deg C, 6 beaker tongs or heat-resistant gloves, 24 wire gauze squares with sintered glass centers.

chemical equation which represents the reaction of zinc metal with copper (II) sulfate. (Hint: zinc sulfate, $ZnSO_4$, is the other product.)

While you are waiting for the zinc to react, write your drawer number in pencil on the etched circle of a 50 mL beaker. Carefully weigh the beaker and record the weight in the data section.

When your instructor indicates the time is up for the above reaction, decant all but 5 mL (looks like a measuring teaspoonful) of the liquid from above the copper and zinc metals in the beaker. Add 10 mL of dilute hydrochloric acid (HCl) **CAUTION: CORROSIVE**, and stir to dissolve any small remaining pieces of zinc metal. Warm the solution gently on the hot plate to speed up the reaction. Hydrogen gas is evolved in the reaction of zinc with the acid so the reaction is complete when bubbles of hydrogen cease to be evolved. (Zinc chloride is also formed in this reaction.) Be sure that the bubbles you see are hydrogen from the reaction and not due to boiling the solution. Write the chemical equation for this reaction on line 9 of your data page.

When the zinc and hydrochloric acid have stopped reacting – there will be no more fizzing – decant the liquid and wash the copper precipitate twice, using 50 mL portions of deionized or distilled water, decanting the wash water each time. Transfer the washed copper metal to the previously weighed 50 mL beaker, keeping the amount of water transferred to a minimum. Decant the water from the beaker. Place the beaker with your product in the oven for drying. This requires about 20–30 minutes, which allows time for you to work on your data page.

When the sample is dry, use beaker tongs or a heat-resistant glove to remove the beaker, and allow it to cool to room temperature on a wire gauze with sintered glass center, to protect the desk top. Accurately weigh it and record the results on your data page.

Calculate the percentage of copper recovered in this experiment, record it on line 5, and show your calculations in the space provided. Place the contents of the beaker containing the copper sample in a container designated by your instructor for proper recycling or disposal.

EXERCISE 15 RECYCLING COPPER – LEARNING TYPES OF REACTIONS

Pre-Lab Questions

Last Name ____________________ **First Name** ________________
Instructor ____________________ **Date** ____________________

1. List at least three everyday uses of copper in our society.

2. Name two acids used in today's lab.

3. Name a base used in today's lab.

4. Explain the precautions for working safely with acids and bases in the lab.

5. How many forms of copper are involved in this experiment?

6. True or False: The reaction of copper with concentrated nitric acid must be done in a fume hood.

7. True or False: It is okay to put the final copper product in the trash.

8. Explain the need for recycling valuable metals such as copper.

EXERCISE 15 RECYCLING COPPER – LEARNING TYPES OF REACTIONS

Data

Last Name ____________________ **First Name** ________________
Instructor ____________________ **Date** ____________________

1. Initial weight of copper wire: ____________________g

2. Initial weight of 50 mL beaker: ____________________g

3. Weight of beaker with recovered copper: ________________ g

4. Weight of recovered copper (#3–#2): ____________________g

5. Percent yield. $\frac{\text{Final weight of Copper (line 4)}}{\text{Initial weight of Copper (line 1)}} \times 100\% =$ _________%

   Show your calculations

6. Copper (II) hydroxide dehydration reaction

   ____

7. Copper (II) oxide + sulfuric acid reaction

   ____

8. Copper (II) sulfate + zinc reaction

   ____

9. Zinc + hydrochloric acid reaction

   ____

10. Did you recover 100% of the original copper in this experiment? If not, explain what might have caused this.

    ____

    ____

EXERCISE 15 RECYCLING COPPER – LEARNING TYPES OF REACTIONS

Post-Lab Questions

You worked with the compounds below in this experiment. Fill in the blanks with the missing name or formula.

Name	Formula
a. Copper (II) sulfate	__________
b. Zinc (II) sulfate	__________
c. __________	$Cu(NO_3)_2$
d. __________	CuO
e. __________	NO_2
f. Nitric acid	__________
g. Potassium hydroxide	__________
h. __________	$Cu(OH)_2$

EXERCISE 16

Studying Evaporation and Preparing "Canned Heat"

Fuels are currently of intense interest due to environmental concerns.

Figure 16.1 Sterno Brand Canned Heat™ is a commercial name for a cooking fuel, which is useful when camping.

INTRODUCTION

The related topics of "Canned Heat," commercially called Sterno™, and "Evaporation" are paired in this study. Evaporation, a common event, is the conversion of liquid to vapor even though the liquid is not at its boiling point.

DOI: 10.1201/9781003006565-19

Evaporation can be useful, e.g., some people like to hang their laundry outside on a line to dry, saving the electricity a dryer would use. If we leave water in an open container or if water remains on the driveway from a recent rain, both of which cases are other examples of an **open system**, the water will ultiatimately evaporate. If the evaporation occurs in a closed container which is a **closed system**, an equilibrium occurs such that the rate of evaporation equals the rate of condensation. An equilibrium occurs when the rate of a forward reaction, such as liquid → vapor equals the rate of the reverse reaction, vapor → liquid. Concentrations of vapor and liquid are then constant, although probably not equal to each other. In this case the pressure of the gas phase is the **vapor pressure**, which is a physical constant. Liquids like the octane in gasoline have high vapor pressures and evaporate readily, which can set off a series of environmentally hazardous reactions in the atmosphere. Liquids like glycerine have low vapor pressures and their evaporation rates are quite slow. To decrease evaporation from water in reservoirs, cetyl alcohol $CH_3(CH_2)_{14}CH_2OH$ is sometimes added as it forms a protective film over the water's surface. Products such as Sterno™ provide a useful heat source and, of course, heat hastens evaporation by providing energy for molecules in the liquid phase to transition to the vapor state.

Fuels are currently of intense interest due to environmental concerns. Fossil fuels such as petroleum are non-renewable energy sources which means that once they have all been used, there will be no more! To extend supplies of petroleum, measures such as the addition of ethanol (aka ethyl alcohol or CH_3CH_2OH) to gasoline are used. When ethyl alcohol has been added to gasoline, the resulting product can be called "gasohol." There are also programs to use ethanol in place of gasoline in internal combustion engines. The ethanol can come from fermentation of plants such as corn and is thus renewable, and it reduces petroleum depletion. Ethanol may also be produced synthetically (with chemicals instead of corn or other biomass). Methods of stretching our fossil fuel supply include conservation as well as the use of additives. Naturally, we need to plan ahead and look to alternative energy sources. These include biomass, hydro-electric, hydrogen, tidal, solar, geothermal, and, yes, nuclear.

During this lab exercise, we will use ethyl alcohol to make a gel. A gel is a form of **colloid**. Living tissue is colloidal, so colloids are extremely important. Colloids have properties in between those of solutions which are clear mixtures, and suspensions which must be constantly shaken or stirred to prevent solid material from sinking to the bottom. Rubbing alcohol is an example of a solution and muddy water is an example of a suspension. Jello™ and jelly are examples of gels; however the gel prepared in this lab is of course not edible. It is flammable and will burn readily. The alcohol will not evaporate as rapidly from the gel as from the liquid state. The semi-solid product made in this exercise is similar to Sterno™.

In this lab, we will examine different factors that affect the rate of evaporation.

INSTRUCTIONS

Safety. Put on your PPE.

Supplies for 24 students or teams of students. 6–12 balances decigram or better, 24 weighing boats, 100 g calcium acetate, deionized/distilled water, 25 mL 6 M sodium hydroxide solution (CAUTION CORROSIVE) dropper bottle recommended, 24 50 mL beakers, 25 mL phenolphthalein indicator solution dropper bottle recommended, paper towels, matches, 1200 mL denatured ethyl alcohol, 24 50 mL or 100 ml graduated cylinders, 24 watch glasses, 24 timing devices, 12–24 aspiration hoses, water aspirators, 24 glass funnels, 24 ring stands, 12 rulers, 24 graduated 1 mL Beral pipettes.

EXERCISE 16 STUDYING EVAPORATION AND PREPARING "CANNED HEAT"

Pre-Lab Questions

Last Name ____________________ **First Name** ____________________
Instructor ____________________ **Date** ____________________

1. Name five alternative sources of energy to replace fossil fuel usage.

2. State the environmental advantage to adding ethyl alcohol to gasoline.

3. Describe the process of evaporation.

4. Name the physical phenomenon which takes place when evaporation occurs in a closed container.

5. Describe the effect of temperature on evaporation.

6. Name the chemicals used in this experiment to form the "canned heat" gel. Write their chemical formulas. Hint: Phenolphthalein's formula is $C_{20}H_{14}O_4$.

7. Explain how you are supposed to dispose of the gel made in this experiment.

8. Name the four different methods used to observe the evaporation of ethanol in this experiment.

EXERCISE 16 STUDYING EVAPORATION AND PREPARING "CANNED HEAT"

Last Name ______________________ **First Name** ________________
Instructor ______________________ **Date** ______________________

Procedure
Note: To save time, set up Part II, step 1 first.

I. **Part I. "Canned Heat" Caution: Ethyl alcohol is flammable so do not light any Bunsen burners.**

1. Measure 8 mL of deionized or distilled water in a graduated cylinder and pour it into a 50 mL beaker. Weigh 3.2 g of calcium acetate in a weighing boat on the laboratory balance and transfer the compound to the water in the beaker. Stir carefully to dissolve all or most of the solid. Be patient! Carefully add eight drops of 6 M sodium hydroxide (NaOH) and stir to mix in the sodium hydroxide. **CAUTION! Sodium hydroxide is corrosive. Avoid eye and skin contact.**
2. Add eight drops of phenolphthalein indicator solution to a 150 mL beaker. Use a 50 mL or 100 mL graduated cylinder to measure 40 mL of ethyl alcohol (ethanol) and add it to the beaker containing the phenolphthalein.
3. While stirring slowly, add the alcohol mixture to the calcium acetate solution. The gel should form immediately. If this does not happen, pour the mixture in the empty alcohol beaker. You should now see the gel. Your instructor may want to initial your lab report sheet after he or she has seen the gel.
4. Prepare a wet paper towel to be used in step 5.
5. Ignite the gel with a match and verify that there is a flame - look carefully as it may not be obvious . Extinguish the flame by placing the wet paper towel from Step 4 over the beaker. A watch glass could be used instead of the wet paper towel, as covering the beaker prevents oxygen from the air from supporting the combustion. Be sure to extinguish the flame and to cover the beaker to prevent evaporation of ethyl alcohol which has a high vapor pressure.
6. Keep your alcohol gel for part II, step 4.

II. **Part II. Evaporation** have your timing device ready to use as soon as the ethanol is poured on the watch glass. Place the watch glass in a part of the lab which is not subject to air currents such as from an air conditioning vent.

1. Add 0.5 mL of ethyl alcohol to a watch glass in an area of the lab that is not drafty. Record the time. Watch the alcohol and record the time needed for complete evaporation. (Useful hint: 0.5 mL is approximately 10 drops.)
2. Repeat Step 1 while waving your lab book back and forth over the watch glass. (Be careful not to hit anyone nearby with your lab book.)
3. Connect the aspiration hose to the water aspirator and a glass funnel. Turn on the water aspirator and place the funnel on the watch glass, which has

been prepared as in Step 1 with 0.5 mL of ethanol. Immediately begin timing the evaporation and continue until you can see through the clear glass funnel that the alcohol is completely evaporated.

4. Place the beaker containing the gel from Part 1 on the base of a ringstand. Adjust the ring so it is about 8 cm from the alcohol gel. Place a watch glass with 0.5 mL of ethyl alcohol on the ring, light the gel and allow the alcohol in the watch glass to evaporate completely. Record the time required.

When you have completed the entire experiment, put your calcium acetate-ethyl alcohol gel in the waste container designated by your instructor.

EXERCISE 16 STUDYING EVAPORATION AND PREPARING "CANNED HEAT"

Data

Last Name ______________________ **First Name** ____________________
Instructor ______________________ **Date** __________________________

I. "Canned Heat" Gel Preparation
Verification by instructor or designated teaching assistant ______
II. Evaporation

Sample Conditions	Ending Time	Initial Time	Total Time (sec)
1 Normal	________	________	______
2 Addition of Air Currents	________	________	______
3 Addition of Vacuum	________	________	______
4 Addition of Heat	________	________	______

EXERCISE 16 STUDYING EVAPORATION AND PREPARING "CANNED HEAT"

Post-Lab Questions

Last Name ____________________ **First Name** ________________
Instructor _____________________ **Date** _____________________

1. What two chemicals formed a color for the gel?

2. What two chemicals formed the gel?

3. True or False: All forms of energy production have the same impact on the environment.____
4. True or False: When ethyl alcohol is burned, it will always burn at the same rate even if another material is mixed with it.

   ____

5. Evaporation occurs when molecules ___________________ from a liquid.

6. Rate the samples from 1-4 as to rate of evaporation, with 4 being the most rapid. _____Normal ______Added Air Currents _________Added Vacuum ______Added Heat

7. What is one factor that can increase the evaporation rate of a liquid?

8. Based on the results of this experiment, which do you think has the higher vapor pressure, ethyl alcohol or water?

9. True or False: It is both safe and legal to dispose of the gel and used gloves from this experiment in the trash. _________________ Explain your answer.

   ____

   _____________________ Adapted from Hassell, Alton C., Marshall, Paula, and Hill, John, Chemical Investigations for Changing Times, 7th ed., Prentice-Hall, Inc., Englewood Cliffs, NJ, 1995, 117–119.

EXERCISE 17

Can We Get Energy from Acid-Base Reactions? Exploring Thermochemistry

While energy is often omitted from a written balanced chemical equation, it is a key factor in the process.

Figure 17.1 Burning candles exemplify exothermic reactions.

DOI: 10.1201/9781003006565-20

INTRODUCTION

A current concern is determining the best use of **energy** for our planet. Energy is defined as the ability to do work. Vehicles, light bulbs, and more are being studied and modified – to save energy! This lab measures energy. During a chemical reaction, energy can be involved in breaking bonds in reactants and forming new bonds in the products. The net energy change in the process is called the **heat of reaction**. That energy difference is usually observed as a temperature change, although something more dramatic like a fire can occur, e.g., when candles (Figure 17.1) or bonfires are lit. Such a reaction which produces energy is called **exothermic**, while baking bread requires energy and is termed **endothermic**.

We can define a **system** as a portion of the universe under study (e.g., a sparkler being lit) and the **surroundings** as the rest of the universe.

The heat of a reaction can be measured with a calorimeter. The energy change from the reaction changes the temperature of the water in the calorimeter. By subtracting the initial temperature of the water from the final temperature of the water, we can determine the heat of the reaction being studied. The calorimeter is constructed as an **isolated system** with insulation so it does not lose heat to the surroundings or gain heat from the surroundings, nor is matter transferred in or out. By determining the amount of heat the water and the calorimeter absorb from the reaction, we can calculate the heat of the reaction.

The amount of heat transferred per mole of limiting reactant, when a reaction is carried out at constant pressure, is known as the **enthalpy change** (ΔH) or **heat of reaction** and is described mathematically as

$$\Delta H_{rxn} = H_{products} - H_{reactants} \quad (17\text{-}1)$$

If the enthalpy change of the reaction is negative, an exothermic reaction is indicated. On the other hand, if the enthalpy change of the reaction is positive, the reaction is termed endothermic.

$\Delta H_{rxn} < 0$ indicates an exothermic reaction.
$\Delta H_{rxn} > 0$ indicates an endothermic reaction.

The heat released (or gained) by the system is equal to the heat gained (or released) by the surroundings. This is due to the First Law of Thermodynamics, which states that energy cannot be created or destroyed. We can show this mathematically as: ΔHsystem + ΔHsurroundings = 0 (17-2)

or in this case:

$\Delta H_{reaction} + \Delta H_{solution} = 0$ (17-3)

$$\Delta H_{reaction} = -\Delta H_{solution}$$

To calculate the heat of a reaction, if we know the number of moles of reactant (n), and its heat capacity (capacity for storing heat) C_p, and the temperature change of the solution ΔT, we can use equation (17-4).

$$\Delta H = nC_p\Delta T \quad (17\text{-}4)$$

Substances with large heat capacities, such as water, can effectively "store" energy, making them more resistant to temperature change. This can be a great advantage for the environment as well as for the different forms of life. In other words, our oceans, lakes, etc. help maintain the Earth's temperature at more constant values; and the water content of life forms assists them in maintaining life-sustaining temperatures. It is a current concern that excessive burning of fuels, etc., can override this safeguard, resulting in climate change.

The heat of the reaction can be calculated by combining the amount of heat the calorimeter components and the water have absorbed. A calibration experiment is usually done to simplify the process. This is done by running a reaction for which we already know the heat of reaction, which allows us to calculate the amount of heat our calorimeter will absorb. Every calorimeter can be different in this respect, so they each need calibrating, even though they may look the same. Then in an actual experiment, the amount of heat absorbed by the calorimeter is known and can be used to calculate the experimental heat of reaction.

INSTRUCTIONS

Safety. Put on your PPE.

Supplies for 24 students or teams of students. [Working in groups of two is recommended to allow one student to record data while the other student stirs the samples and reads the thermometer.] 24 8-12 ounce Styrofoam™ cups, 24 -10–100°C thermometers (recommend non-mercury type), 800 mL 3.0 M HCl CAUTION: CORROSIVE, 24 thermometer clamps, 24 ring stands with thermometer clamps or other small clamps, 800 mL 3.0 M NaOH CAUTION: CORROSIVE, 24 100 mL graduated cylinders, 24 timing devices, deionized/distilled water, 250 g $Ba(OH)_2 \cdot 8H_2O$ CAUTION: TOXIC, 130 g NH_4NO_3, 24 250 mL beakers, 2.5 L 10% w/v Na_2SO_4 solution, 24 glass stirring rods, 12–24 Buchner funnels with suction flasks, vacuum tubing, water aspirators, filter paper to fit funnels, 24 weighing boats, 6-24 balances decigram or better, 24 glass stirring rods with rubber policemen, 24 200 mm test tubes with stoppers, deionized/distilled water.

PART I. NEUTRALIZATION REACTION OF HYDOCHLORIC ACID AND SODIUM HYDROXIDE

$HCl + NaOH \longrightarrow NaCl + H_2O$ _______ calories/mole (17-5)

1. Obtain a Styrofoam™ cup and a thermometer. Stand the thermometer up in a beaker to stabilize it. Laying thermometers flat on the desk is not allowed as it can cause the liquid in the thermometer to separate.
2. In a graduated cylinder, measure carefully 25.0 mL of 3.0 M HCl **CAUTION: CORROSIVE** - avoid eye and skin contact and pour it into the Styrofoam™ cup.
   CAUTION: Hydrochloric acid (HCl) is corrosive and can cause burns to the eyes or to the skin! If splashed in your eye, immediately use the eye wash fountain for 15 minutes to remove the HCl from your eyes. If spilled on skin, rinse copiously with water and notify your instructor.
3. Using a small clamp attached to a ring stand, support your thermometer in the cup.
4. Record the temperature of the 25.0 mL of HCl in the cup. This is the initial temperature.
5. Using a clean graduated cylinder, measure 25.0 mL of 3 M NaOH.
   CAUTION: Sodium Hydroxide (NaOH) is corrosive and can cause burns to the eyes and skin. If spilled, wash with large amounts of water until the "slimy" feel to your skin is gone. If splashed in your eye, immediately use the eye wash fountain for 15 minutes to remove the NaOH from your eyes. Notify your instructor.
6. Carefully pour the 25.0 mL of 3 M NaOH into the Styrofoam cup containing the HCl.
7. Carefully swirl the cup to mix the contents without spilling.
8. Using your supported thermometer, record the temperature of the resulting mixture one minute after mixing. Continue to record the temperature at one-minute intervals until the temperature levels off or remains constant. Use the highest temperature reading in your calculations.
9. The resulting solution can be disposed of by pouring down the sink, if your instructor approves.
10. Rinse your Styrofoam™ cup with water and allow it to dry so it can be reused.

PART II. REACTION OF BARIUM HYDROXIDE WITH AMMONIUM NITRATE

$Ba(OH)_2 + 2\ NH_4NO_3 \longrightarrow Ba(NO_3)_2 + 2\ NH_3 + 2H_2O +$ __________ calories/mole (17-6)

1. Measure 50.0 mL of deionized/distilled water in a graduated cylinder and add it to the Styrofoam™ cup.
2. Obtain a clean, dry 200 mm test tube with stopper to fit
3. Weigh a weighing boat on the balance; then add sufficient $Ba(OH)_2{\bullet}8H_2O$ **CAUTION: TOXIC** to *increase* the weight by 10.0 g
4. Clamp the thermometer securely in the ring stand and measure the temperature of the water in the Styrofoam™ cup, allowing time for the reading to stabilize.
5. Weigh a weighing boat on the balance; then add sufficient ammonium nitrate (NH_4NO_3) to increase the weight by 5.0 g.
6. Add the NH_4NO_3 to the test tube containing the $Ba(OH)_2{\cdot}8H_2O$. Immediately stopper the test tube and shake until the solid in the tube becomes a liquid.
7. Place the test tube in the Styrofoam cup containing the water. Support the test tube with a clamp attached to the ring stand.
8. Record the temperature after one minute and every minute after until the temperature no longer changes. Use the *lowest* temperature reading in your calculations.

Save the test tube and its contents for Part III.

Part III Synthesis of a Diagnostic Medical Compound

$Ba(OH)_2 + Na_2SO_4 \rightarrow BaSO_4 + 2NaOH$ (17-7)

The $BaSO_4$ (barium sulfate) of a very pure formulation – not that synthesized here which has not been purified – is given to patients having some X-ray diagnostic tests. Never consume any chemicals in the lab, or remove any chemicals from the lab.

1. Pour the contents of the test tube saved from Part II into a 250 mL beaker.
2. Measure 100 mL of 10% sodium sulfate solution.
3. Use some of the 100 mL 10% sodium sulfate solution to rinse your test tube. Pour the rinse into the 250 mL beaker. Pour the remainder of the 10% sodium sulfate solution into the beaker.
4. Stir the resulting solution and allow the barium sulfate salt ($BaSO_4$) to settle.
5. Set up a Buchner funnel and suction flask. Place a piece of filter paper in the funnel so that it covers all of the holes and lies flat. Attach the vacuum hose to the side arm of the suction flask and to the water aspirator. Turn on the water and pour off the liquid from your test tube (in other words, decant the liquid) onto the flask's filter paper. Transfer the solids from the test tube into the funnel using a glass rod with a rubber policeman. Rinse the beaker with a small amount of deionized/distilled water to remove as much of the solids as possible and discard solids including the filter paper into the designated waste container.

Adapted from: Hassell, C. Alton, Marshall, Paula, and Hill, John W., *Chemical Investigations for Changing Times Seventh Edition*, Prentice-Hall, Inc., Englewood Cliffs, NJ, 1995, 107–109.

EXERCISE 17 CAN WE GET ENERGY FROM ACID-BASE REACTIONS? EXPLORING THERMOCHEMISTRY

Data

Last Name ____________________ **First Name** ________________

Instructor ____________________ **Date** ______________________

Part I Reaction of Hydrochloric Acid and Sodium Hydroxide

Initial Temperature ______°C

Temperature at 1 minute _______°C

Temperature at 2 minutes ______°C

Final Temperature _______°C

Temperature at 3 minutes ______°C

Difference in Temperature:

Temperature at 4 minutes ______°C

$T_{\text{final}} - T_{\text{initial}} = \Delta T =$ _______°C

Temperature at 5 minutes ______°C

$\Delta H = -\Delta T°\text{C} \times 50\ \text{mL} \times 1.00\ \frac{\text{cal}}{\text{mL}\,°\text{C}} =$ ______calories ÷ 0.075 moles

$= \underline{\hspace{2em}}\ \frac{\text{calories}}{\text{mole}}$

The reaction is ____________________ (exothermic, endothermic).

Part II Reaction of Barium Hydroxide and Ammonium Nitrate

Initial Temperature of water _____°C

Temperature of Water at

1 minute ________°C

Final Temperature ______°C

2 minutes _______°C

Difference in temperature:

3 minutes _______°C

$T_{\text{final}} - T_{\text{initial}} = \Delta T =$ _______°C

4 minutes _______°C

5 minutes _______°C

$\Delta H = -\Delta T$_____°C × 50 mL × 1.00 $\frac{\text{cal}}{\text{mL}\,°\text{C}}$ = ______calories ÷ 0.031 moles

$= \Delta H \frac{\text{calories}}{\text{mole}}$

The reaction is ________ (exothermic, endothermic).

EXERCISE 18

Are Electrolytes Changing Our Waters? Identifying Electrolytes with Conductivity Measurements

Marine life is endangered when the electrolyte concentrations of natural waters are too high, or too low.

Figure 18.1 Athletes often consume electrolyte-enriched beverages.

INTRODUCTION

Electrolytes! We all need them, and in the correct concentration. Various products such as that shown in Figure 18.1 are marketed with advertising that they replace missing electrolytes in the body. Similarly, our lakes and

DOI: 10.1201/9781003006565-21

streams, rivers and oceans, need the proper electrolyte concentrations. Like Goldilocks' porridge, which was neither too hot nor too cold, for health reasons, electrolyte concentrations cannot be too high or too low. There is quite a difference in the salt concentration of freshwater and seawater. Did you ever manage a salt water aquarium? If so, you have considerable experience in this area.

Marine life is endangered when the electrolyte concentrations of natural waters are excessive or deficient. In a **hypotonic** solution, the electrolyte concentration of the surrounding water is too low, and excess water will enter the marine organisms, leading to cell enlargement and possible bursting. In contrast, in a **hypertonic** solution, the electrolyte concentration of the surrounding water is too high, and water will leave the organism and cells will shrink, leading to dehydration. That is why we cannot survive drinking salty water. Simply stated, electrolytes are compounds which form **ions** in solution, and these ions travel through the solvent. The word *ion* comes from a Greek word meaning "go," because ions *go* toward electrodes with opposite charge. For example, in a salt water solution, the Na^+ ions will travel toward the negatively charged electrode, and the Cl^- ions will travel toward the positively charged electrode. This effect can be quantified with a voltmeter, or simply demonstrated with a light bulb, as shown below in Figure 18.2. **Strong electrolytes** will produce many ions traveling between the electrodes, resulting in a bright light. This is illustrated by the following equation, which represents 100% of the NaCl units forming ions. The ions must be *mobile* to conduct electricity, i.e., the compound must be dissolved in water – or melted. No conduction would be observed with electrodes applied to a block of *dry* sodium chloride.

$$NaCl(s) \rightarrow Na^+(aq) + Cl^-(aq)$$

Weak electrolytes such as acetic acid will produce few ions, with slight conduction of electricity, causing only a dim light. The following equation with the *reversible* arrow illustrates this.

$$HC_2H_3O_2(l) \rightleftarrows H^+(aq) + (C_2H_3O_2)^-(aq)$$

Weak electrolytes, however, may react with each other to form strong electrolytes, as in the case of concentrated ammonia and glacial (without water or anhydrous) acetic acid, as shown below.

$$CH_3COOH + NH_3 \rightarrow NH_4^+(aq) + CH_3COO^-(aq)$$

And **non-electrolytes** such as sucrose (table sugar) will produce no ions at all, resulting in no conduction of electricity and thus no light. Non-

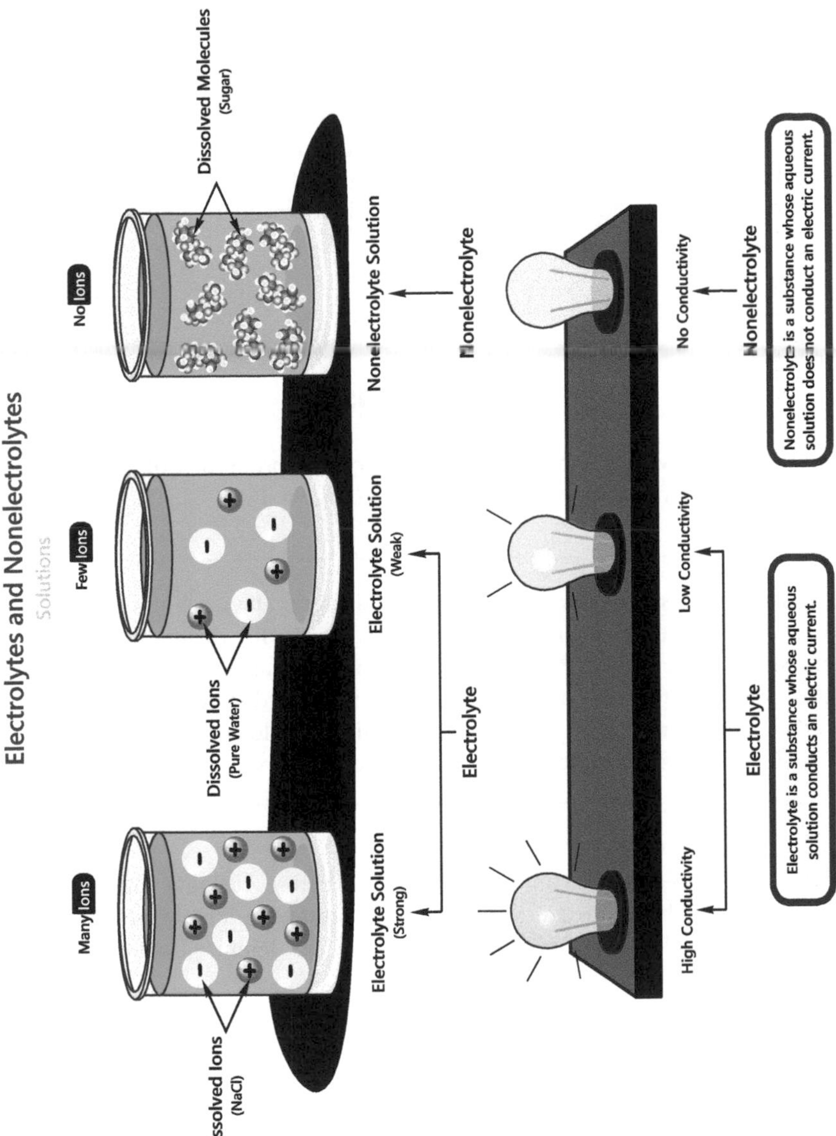

Figure 18.2 Conductivity device with electrolytes and nonelectrolytes.

electrolytes belong to the **covalent** class of compounds, which bond by **sharing electrons**. Ionic compounds form by **donating and receiving electrons**. Your instructor will provide a demonstration of conductivity effects for you to record on your data sheet.

In summary, the two main classes of chemical compounds are ionic and covalent. Ionic compounds tend to form between elements on opposite sides of the periodic table, i.e., between metals on the left, and non-metals on the right. Covalent compounds tend to form between different non-metals. As examples, Li_2O lithium oxide is an ionic compound (electrolyte if melted or dissolved in water) and CO_2 carbon dioxide is a covalent compound (non-electrolyte).

INSTRUCTIONS

Your instructor will demonstrate the conductivity properties of a number of solutions using a device similar to that shown in Figure 18.2.

Supplies for classroom demonstration. Conductivity device with light bulb; twelve 250 mL beakers; large beaker, e.g., 2000 mL for rinsing electrodes; table sugar; table salt; concentrated ammonia; glacial acetic acid; 0.1 M NaOH solution; 0.1 M HCl solution; deionized or distilled water; paper towels; PPE for instructor; transparent lab safety shield for demonstrations. CAUTION: Acids and bases in this exercise are corrosive.

CAUTION: The apparatus must be unplugged before cleaning the electrodes. The electrodes must be cleaned after each sample is tested, before the next sample is studied. The electrodes are to be dipped in distilled or deionized water and dried with a paper towel.

EXERCISE 18 ARE ELECTROLYTES CHANGING OUR WATERS? IDENTIFYING ELECTROLYTES WITH CONDUCTIVITY

Pre-Lab Questions

Last Name ____________________ **First Name** ____________________
Instructor ____________________ **Date** ____________________

1. The two main classes of chemical compounds are

2. For the light bulb conductivity experiment described above, weak electrolytes, when dissolved in water, produce
   a. much electrical conductivity with a bright light
   b. much electrical conductivity with a dim light
   c. only slight electrical conductivity with a bright light
   d. only slight electrical conductivity with a dim light
3. A cucumber placed in concentrated salt water will
   a. expand by absorbing water
   b. shrink by losing water
   c. remain the same
   d. disintegrate
4. Select the compound that is not a strong electrolyte.
   a. KF (potassium fluoride)
   b. NaCl (sodium chloride)
   c. LiBr (lithium bromide)
   d. $C_6H_{12}O_6$ (dextrose)
5. Select the sample that would provide the greatest amount of electrical conductivity.
   a. Concentrated NaCl solution
   b. Solid ice cream salt
   c. Dilute NaCl solution
   d. Sugar water

EXERCISE 18 ARE ELECTROLYTES THREATENING OUR WATERS? IDENTIFYING ELECTROLYTES WITH CONDUCTIVITY

Data

Last Name ____________________ **First Name** ____________________
Instructor ____________________ **Date** ____________________

Complete Table 18.1 based on the instructor's demonstration.

Table 18.1 **Conductivity Test Results**

Sample	Type of Light (Bright, Dim, or None?)	Conclusion (Strong Electrolyte, Weak Electrolyte, or Non-Electrolyte?)
Distilled or deionized water		
Tap water		
0.1 M NaOH (sodium hydroxide) solution		
0.1 M HCl (hydrochloric acid) solution		
Concentrated NH_3 (ammonia) solution		
Glacial CH_3COOH (acetic acid)		
Solid NaCl (sodium chloride)		
Solid NaCl (sodium chloride) with distilled or deionized water added		
0.1 M $C_{12}H_{22}O_{11}$ (sucrose or table sugar) solution		
0.1 M NH_3 (ammonia) solution with an equal amount of 0.1 M $HC_2H_3O_2$ (acetic acid) solution added		
Concentrated Ammonia with Glacial Acetic Acid		

EXERCISE 18 ARE ELECTROLYTES CHANGING OUR WATERS? IDENTIFYING ELECTROLYTES WITH CONDUCTIVITY

Post-Lab Questions

Last Name ____________________ **First Name** ____________________
Instructor ____________________ **Date** ____________________

1. How could you distinguish between sugar and salt based on the methods of this experiment?

2. Explain why neither concentrated ammonia nor glacial acetic acid conducted current.

3. Explain why a mixture of concentrated ammonia and glacial acetic acid conducted current.

4. Write a balanced chemical equation for the reaction in Question *3 above.

5. Explain why distilled water does not conduct a current, but tap water does so, weakly.

6. Explain why solid NaCl does not conduct a current, but aqueous or molten NaCl does conduct a current.

EXERCISE 19

Radiation, an Invisible Pollutant – Measurements and Calculations

While we cannot avoid all radiation, we do have some control over our exposure to it.

Figure 19.1 Applications of nuclear radiation in our environment.

DOI: 10.1201/9781003006565-22

INTRODUCTION

Around 1807 the brilliant English chemist John Dalton formulated the Atomic Theory, much of which we still use today; however, the limited experimental information available at that time did not allow Dalton to recognize that some atoms simply were not stable. The discovery that these atoms are busily splitting apart all around us was left to scientists such as Marie and Pierre Curie almost 100 years later. This discovery has provided us with both benefits and concerns. Today nuclear radiation, as shown in Figure 19.1, is used for energy, medicine, and other sectors of our society. This lab focuses on the nucleus of the atom, not the electrons which control chemical reactions. The nucleus contains protons which have a positive charge, and neutrons which are electrically neutral (zero charge).

Unstable atoms create the **background** radiation which is just naturally present in the environment - even our bodies contain radioactive **isotopes** although in very small amounts. Isotopes are different forms of the same element. While all isotopes of an element contain the same number of protons, e.g., carbon atoms always have exactly six protons, the number of neutrons can vary. Carbon-12 and carbon-13 with six and seven neutrons, respectively, are not radioactive, but the rarer isotope carbon-14, with 8 neutrons, is radioactive. Figure 19.2 show common sources of radiation exposure for US residents.

While we cannot avoid all radiation, we do have some control over our exposure to it. As an example, Magnetic Resonance Imaging (MRI) studies have replaced X-rays for some diagnoses, as a safer alternative. This lab allows us to estimate our annual exposure to radiation in units of millirem which are explained below.

Units: Millirem (mrem) per year. Rem means roentgen equivalent man, i.e., the radiation amount that has the same effect on a man or woman as 1 roentgen of X-ray.

INSTRUCTIONS

Supplies for 24 students. Handheld radiation meter (1), Brazil nuts, bananas, cigarettes, 12-24 calculators.

Your instructor may use a handheld radiation meter to demonstrate the background radiation in your classroom, as well as the radiation given off by Brazil nuts and bananas, both of which have higher radiation levels than many other foods. While these items are generally considered safe, it is interesting to see if any radiation counts above background can be detected for them with the device used in this lab. Bananas, e.g., contain K-40, a radioactive isotope of potassium, and Brazil nuts contain Ra-226 and

Ra-228), isotopes of radium. Polonium-210 is in cigarettes, which are unsafe regardless of the presence or absence of radioactivity. Record this information in Table 19.1.

Next, complete Table 19.2, which shows our common exposure to various forms of radiation, and total your exposure. For example, in the first line, Cosmic radiation, you would list 26. The second line is for the altitude at which you live. As an example, if you live at an altitude of 130 feet above sea level, you would record a "130" in the second line of the second column, and a "2" on the third line under "Your Values," and skip the next four lines. Please omit medical/dental procedures to protect your privacy. Please star the source(s) to which we can control our exposure.

EXERCISE 19 RADIATION, AN INVISIBLE POLLUTANT - MEASUREMENTS AND CALCULATIONS

Pre-Lab Questions

Last Name ____________________ **First Name** ____________________
Instructor ____________________ **Date** ____________________

1. Give the term for the radiation which is naturally around us in the environment.

   ____

2. True or False: John Dalton's Atomic Theory stated that atoms can be unstable.

   ____

3. True or False: Isotopes of the same element are all equally safe.

   ____

4. Isotopes of the same element have the same number of ______________ but a different number of.

   ____

5. Name a diagnostic medical procedure which does not involve nuclear radiation and which has replaced X-rays for some tests.

   ____

6. In this lab we will estimate our exposure to nuclear radiation in units of.

   ____

EXERCISE 19 RADIATION, AN INVISIBLE POLLUTANT - MEASUREMENTS AND CALCULATIONS

Data

Last Name ____________________ **First Name** ____________________

Instructor ____________________ **Date** ____________________

Table 19.1 Background and Miscellaneous Radiation Tests

Areas of Classroom Tested	Counts per Minute
Background Radiation	
Area 1, Trial 1	
Area 1, Trial 2	
Average	
Area 2, Trial 1	
Area 2, Trial 2	
Average	
Area 3, Trial 1	
Area 3, Trial 2	
Average	
Some Naturally Radioactive Objects	
Bananas, Trial 1	
Bananas, Trial 2	
Average	
Brazil nuts, Trial 1	
Brazil nuts, Trial 2	
Average	
Cigarettes (unlit) Trial 1	
Cigarettes (unlit) Trial 2	
Average	

Table 19.2 Radiation Sources and Their Amounts in the Environment

Radiation Source	Annual Radiation Dose mrem	Your Values
Cosmic radiation at sea level from outer space	26	
Your elevation from sea level	________	
Up to 1000 feet	2	
1000–2000 feet	5	
2000–3000 feet	9	
3000–4000 feet	21	
4000–5000 feet	29	
Terrestrial radiation from ground if in a state bordering the Pacific or Gulf Coast	23	
Food and water	40	
Air (radon)	200	
Reside in a stone, brick, or concrete building	7	
Weapons test fallout	Less than 1	
Miles of travel by jet plane	1 mrem per 1000 miles	
Porcelain crowns or false teeth	0.07 mrem	
Gas lantern mantles used for camping	0.003	
Luggage inspection at airports	0.002	
Video display terminal use	Less than 1	
Smoke detector	0.008	
Plutonium-powered cardiac pacemaker	100	
Diagnostic X-rays	40	
Nuclear medicine procedures such as thyroid scans	14	
Live within 50 miles of a nuclear power plant with a pressurized water reactor	0.0009	
Live within 50 miles of a coal-fired electrical utility power plant	0.03	
Total		

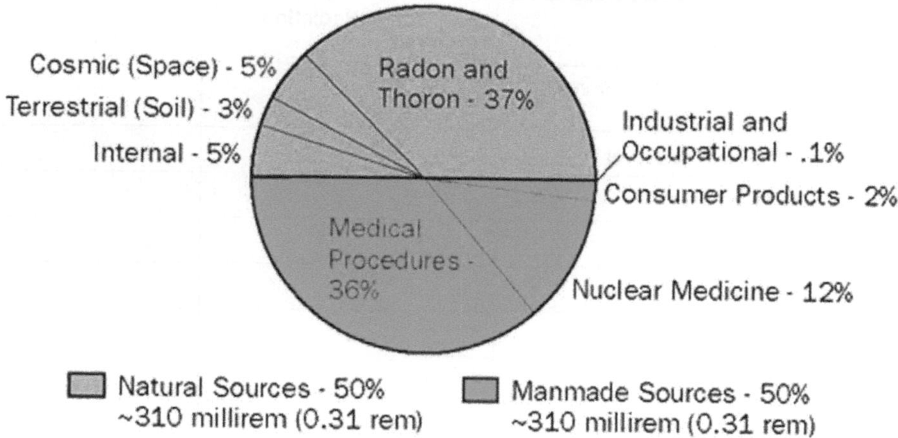

Figure 19.2 Percentages of nuclear radiation exposure in our environment.

BIBLIOGRAPHY

www.nrc.gov/about-nrc/radiation/around-us/calculator.html

EXERCISE 20

Is Extremely Low Frequency Radiation Harmful? A Laboratory Analysis

"Mathematics is the queen of the sciences."

– Carl Friedrich Gauss, German mathematician and physicist

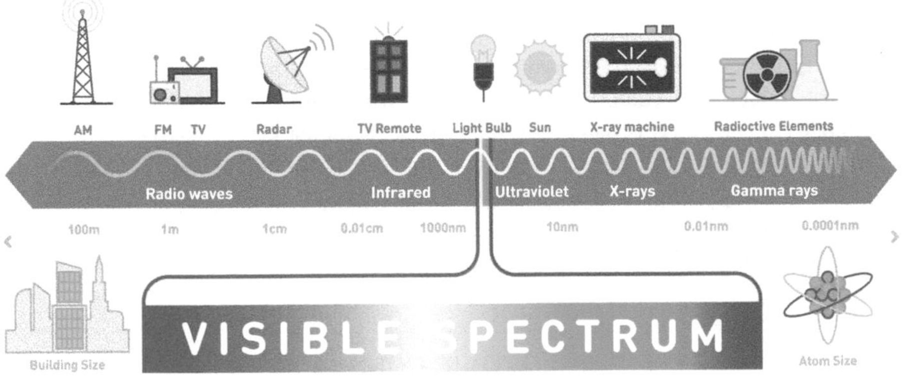

Figure 20.1 Electromagnetic spectrum.

INTRODUCTION

Radiation! Some people shy away from the very word, but a knowledge of science tells us some radiation types are either harmless or can actually promote life. We depend on visible light to see, and sometimes residents of areas with less sunshine, e.g., some coastal areas, are prone to seasonal affective disorder (SAD), a type of depression. Remember the **electromagnetic spectrum**

DOI: 10.1201/9781003006565-23

(EMS), which can range from very low energy **radio frequency rays** to very high energy **gamma rays**. As shown in Figure 20.1, radiation with high energy has high frequency and short wavelength, while low-energy radiation has low frequency but long wavelength. High energy radiation can be both dangerous and helpful; while these rays can damage cells, they are used in medicine to excise brain tumors – and called the "Gamma Knife." This lab focuses on **extremely low frequency (ELF) electromagnetic radiation** which is located to the left of AM radio waves in Figure 20.1.

The electricity we depend on in our daily life causes small, but regular exposure to electrical and magnetic radiation beyond that which occurs naturally in the environment. This additional radiation emanates from the appliances and electronic devices we rely on, and the neighborhood high-voltage power lines which they require. Although conclusive evidence is not available, concern exists that cancer can be more prevalent in areas near power stations and substations, and transmission lines. Studies have probed the possibility that electric blanket use, and the resultant exposure to electromagnetic frequency (EMF), can cause leukemia and breast cancer; however, definitive proof is not available for this either.

In this exercise, we study the unseen world of ELF which is all around us. Various battery-powered devices of the **direct current (DC)** type as well as **alternating current (AC)** items are studied. We investigate various appliances, power lines, and the radiation strength involved, both closeup and at a distance. The ELF meter measures this in units of **milligauss**. A milligauss is 1/1000 of a gauss, which is 1/10,000 of a tesla, all of which are units of **magnetic flux density**. [The gauss is named for the German scientist **Carl Friedrich Gauss**, and the **tesla** for the Serbian-American inventor **Nikola Tesla**.] For comparison, a strong refrigerator magnet has a field strength of about 100 gauss, and MRI instruments used for diagnosing bone fractures and other problems range from 20,000 to 70,000 gauss in field strength.

INSTRUCTIONS

Supplies. ELF meters, metric rulers, electrical appliances and devices.

Use the ELF meter to measure the EMF in milligauss produced by the electric appliances and devices assigned by your instructor; then record the results on your data page. [If the number of meters is limited, your instructor may conduct the lab as a demonstration, possibly with student participation.] Suggestions include cell phones, computers, iPads, calculators, radios, TVs, microwave ovens, electric heating pads, hot plates, and refrigerators. Ensure that the device to be measured is turned on. Hold the ELF meter 3.0 cm from the device, measure, and record the ELF strength. Repeat at a distance of 30.0 cm. If available, measure the ELF of a nearby power line. Stand directly underneath the line and record the field strength. Then move to a distance approximately 10 meters or 33 feet from underneath the line and record the field strength.

EXERCISE 20 IS EXTREMELY LOW FREQUENCY (ELF) RADIATION HARMFUL? A LABORATORY ANALYSIS

Pre-Lab Questions

Last Name ______________________ **First Name** ______________

Instructor ______________________ **Date** __________________

1. A milligauss is a unit of
   a. frequency
   b. energy
   c. magnetic flux density
   d. wavelength

2. Low frequency radiation has
   a. high energy and long wavelength
   b. low energy and long wavelength
   c. low energy and short wavelength
   d. high energy and short wavelength

3. Radiation is found only in the laboratory
   a. True
   b. False

4. ELF stands for.

   ____

5. Which of the following can be used for brain surgery?
   a. Visible light
   b. Microwave radiation
   c. Infrared radiation
   d. Gamma rays

6. Which has the greater magnetic field strength, a refrigerator magnet or an MRI instrument?

   ____

EXERCISE 20 IS EXTREMELY LOW FREQUENCY (ELF) RADIATION HARMFUL? A LABORATORY ANALYSIS

Data

Last Name ______________________ **First Name** ____________________
Instructor ______________________ **Date** ____________________

Appliance/Device	ELF milligauss at 3.0 cm	ELF milligauss at 30.0 cm

Power Line, Directly Below

Power Line, at 10 Meters Distance

EXERCISE 20 IS EXTREMELY LOW FREQUENCY (ELF) RADIATION HARMFUL? A LABORATORY ANALYSIS

Post-Lab Questions

Last Name ________________________ **First Name**____________________
Instructor ________________________ **Date** ________________________

1. List the appliances and devices you studied in order of decreasing ELF values, beginning with the highest and ending with the lowest.

2. Do any of the appliances and devices studied seem to pose a significant risk to you? Consider the time you spend around them as well as the amount of ELF you measured. If so, which stand out as more hazardous?

3. Was there any difference between the readings at 3.0 cm and 30.0 cm for the same object? Explain your results. What do you think the effect of distance is?

4. Did any of your battery-powered (direct current or DC) devices create an ELF? If so, how did its field strength compare to the alternating current or AC-powered appliances?

5. Would you be concerned to live near a power station such as one with lines like those shown in Figure 20.2? Explain.

Figure 20.2 Measurement of extremely low radio frequency near power lines.

EXERCISE 21

Adventures in Organic Chemistry Land

Let us learn to dream, ..., then perhaps we shall find the truth.

– August Kekule, German organic chemist

Figure 21.1 Stamp featuring August Kekule and the structure of benzene which he discovered.

DOI: 10.1201/9781003006565-24

INTRODUCTION

The pesticide DDT (dichlorodiphenyltrichloroethane), the biomass fuel ethanol, petroleum, natural gas, and more are members of a fascinating class of chemicals – organic compounds! The very word *organic* seems to convey the idea of life. Many of the molecules involved in air pollution (e.g., peroxyacetyl nitrate aka PAN) and water pollution (e.g., dimethyl mercury) are organic compounds. These compounds all contain the element carbon which has a unique way of bonding so that it can form many, many compounds – over nine million known is the current estimate! Chemists often look at structures to predict the properties of various organic compounds and to design new pharmaceuticals. The famous German chemist August Kekule solved a structural chemistry problem by dreaming! A stamp produced in his honor is shown in Figure 21.1 Let's explore a few of these compounds by looking at their structures.

INSTRUCTIONS

Your instructor will let you know if you are to use the kit method, the sketch method, or both.

Supplies for 24 students. 12 -24 molecular model kits and/or highlighters, colored pens, paper.

KIT METHOD

Use the kits to construct models of the molecules listed under Sketch Method. Then, use the model to sketch the molecule. Use springs for double bonds, short sticks for bonds to hydrogen, and the longer sticks for all other bonds. Many molecular modeling kits use the following color code: nitrogen = blue; oxygen = red; chlorine = green; hydrogen = yellow or white; and carbon = black. Your instructor or teaching assistant (TA) may want to approve and check off (initial) each model you construct. For online work, take pictures of your molecular models to submit.

SKETCH METHOD

You will need highlighters, or colored pens, etc. Sketch the structures of the following molecules. Be sure to label each with its symbol. Here is an example for the greenhouse gas carbon dioxide. Be sure that you follow the color code above, though, where oxygen atoms are red and carbon atoms are black. You must show the bonds. The double bonds (= in carbon dioxide below) join the atoms to form the molecules, for example, O=C=O.

EXERCISE 21 ADVENTURES IN ORGANIC CHEMISTRY LAND

Last Name ________________________ **First Name** ______________

Instructor ________________________ **Date** ____________________

1. Ethane has the formula C_2H_6 and the condensed formula CH_3CH_3. Draw two black carbon atoms linked together with a single bond or line. Next draw three yellow, smaller hydrogen atoms with a single bond to each carbon atom.

2. Ethylene (aka ethene) has the formula C_2H_4 and the condensed formula CH_2CH_2. Draw two black carbon atoms linked together with two bonds (like =). Next draw two yellow, smaller hydrogen atoms with a single bond to each carbon atom. Ethylene is used in the manufacture of plastics called polyethylene.

3. Acetylene (aka ethyne) has the formula C_2H_2 and the condensed formula CHCH. Draw two black carbon atoms linked together with three lines to represent the bonds. Next draw one yellow, smaller hydrogen atom with a single bond or line to each carbon atom. Acetylene is used by welders to cut metal (!) and to manufacture vinyl plastics.

4. Butane has the formula C_4H_{10} and is an important fuel. There are two isomers (different molecules with this formula).

   First, draw normal or n-butane which has the condensed formula $CH_3CH_2CH_2CH_3$.

   Next, draw iso-butane which has the condensed formula $(CH_3)_2CHCH_3$. All the atoms in the butanes are linked by a single bond or line. Remember to make the carbons black and larger than the smaller, yellow hydrogens.

Normal Butane	**Iso-Butane**

5. Ethanol (also known as beverage alcohol and a gasoline additive!) has the formula C_2H_6O and the condensed formula CH_3CH_2OH.
   First, draw a larger black carbon atom and a bond or line to connect it to a second carbon atom. On the first carbon atom, attach three yellow, smaller hydrogen atoms. On the second carbon atom, attach two yellow, smaller hydrogen atoms; then to the second carbon atom attach a medium-sized red oxygen atom. To the oxygen attach a little hydrogen atom. The OH group is so cool – it causes the molecule to be an alcohol. Yes, there are many alcohols – not just one.

6. Acetic acid (vinegar is usually 5% acetic acid) has the condensed formula CH_3COOH. The COOH group is also cool – it causes the molecule to be an organic (aka carboxylic) acid!

   First, draw a larger black carbon atom and a bond to a second carbon atom. To the first carbon atom, bond three smaller, yellow hydrogen atoms. To the second carbon atom, draw a double bond (like this =) and attach a red oxygen atom. Then, to the second carbon atom also draw a single bond (like this –) and attach a red oxygen atom; to the red oxygen atom, bond a smaller, yellow hydrogen atom.

7. Ethyl acetate belongs to a class of organic molecules called *esters*. Ethyl acetate is in non-acetone nail polish removers. Some of these *esters* smell wonderful and are used in perfumes, candles, etc. Amyl acetate, an *ester*, has the fragrance of bananas and is artificial banana flavoring!

   The condensed formula of ethyl acetate is $CH_3–CH_2–O–(C{=}O)–CH_3$.

   First, draw a larger black carbon atom; then attach three smaller, yellow hydrogen atoms to it; then attach a second, larger black carbon atom to the first carbon atom. Then attach two yellow hydrogens to the second carbon atom. Then attach a red oxygen to the second carbon. Then attach a black carbon to the first oxygen. Then attach an oxygen with a double bond (=) to the third carbon. Last, attach a carbon with three hydrogens to the third carbon.

8. Chloroform has a relatively simple formula $CHCl_3$. It was used as a general anesthetic until safer options became available. After Queen Victoria had two children while being given chloroform, the social elite wanted this anesthesia for their childbirth experiences, too.

   First draw a black carbon. Make four single bods around it. Place three large green chlorine atoms and one smaller yellow hydrogen on those bonds.

9. Methyl amine has the condensed formula CH_3NH_2.

   First, sketch a carbon with three hydrogens around it; then draw a line with a blue nitrogen attached. Draw two yellow hydrogens on the nitrogen, and you are done! Hey, amines can smell like old fish ☹ – no perfume here!

10. Glycine is an – *amino acid*! Amino acids link together in our bodies to make long chains which we call proteins!!

    The condensed formula of glycine is NH_2CH_2COOH.

    First, sketch a medium-sized blue nitrogen with two smaller yellow hydrogens attached. Then draw a line from the nitrogen and draw in a black carbon atom. Add two yellow hydrogens to the carbon atom. Then draw a line from the carbon atom and add another carbon atom. To the carbon, attach a red oxygen with a double bond and a red oxygen with a single bond. To the oxygen with a single bond, add a smaller yellow hydrogen atom.

Here is a picture of a protein molecule in Figure 21.2 They are really *giant* molecules!

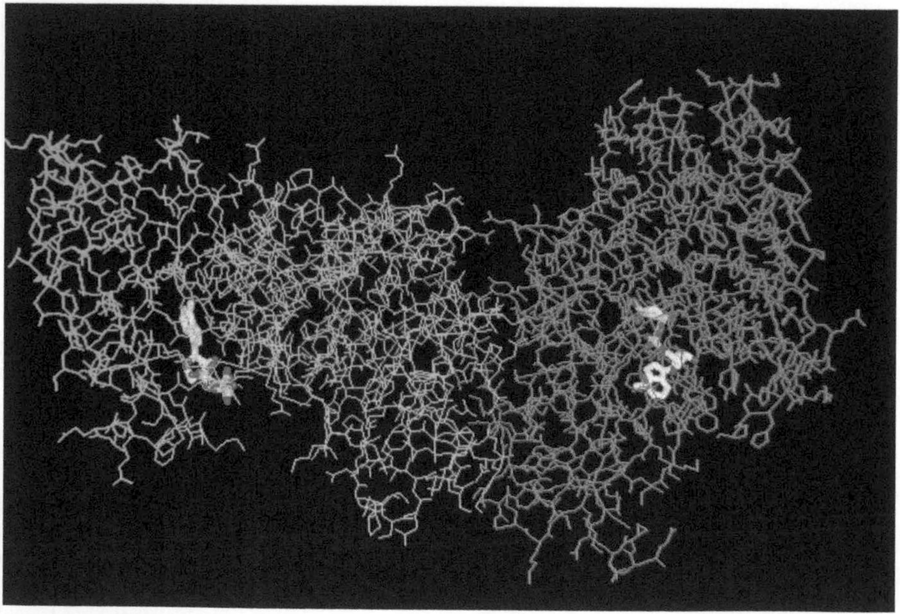

Figure 21.2 Protein.

Before you finish your paper, be sure you have labeled each atom: C for carbon, H for hydrogen, etc. Be sure to draw in the correct number of bonds to show how the molecule is held together. You may use your notebook, if your instructor approves, to create more space to draw in your "molecules." Be sure your submission is neat!

EXERCISE 22

What's in Red Cabbage? Adventures in pH Land

Anthocyanins are responsible for many of the colors we enjoy in the natural environment.

Figure 22.1 Red cabbage is a source of a pH indicator.

DOI: 10.1201/9781003006565-25

INTRODUCTION

Environmental Chemistry must deal with acid rain, soil and water pH, and their effects on life and property. This lab provides experience with acid-base chemistry, anthocyanins, and the pH determination of some common household products and foods such as the red cabbage shown in Figure 22.1.

PART I ANTHOCYANINS

Anthocyanins are responsible for many of the colors we enjoy in the natural environment. Flowers are often blue or red because of anthocyanins. The beautiful fall deciduous forests of New England owe much of their spectacular color to anthocyanins. As the leaves enter senescence (deterioration) with shorter days and cooler weather, chlorophyll begins to decompose and anthocyanins are synthesized. Many of the fruits and vegetables we enjoy are appealing in part because of the color of anthocyanins; the list includes grapes, blueberries, cranberries, raspberries, blackberries, blood orange, apple, eggplant, red onion, red cabbage, and even *blue* corn! The anthocyanins are antioxidants and are studied by chemists to determine their health benefits.

The structure of anthocyanin is shown in Figure 22.2. It can be modified by adding various chemical groups so that over 500 anthocyanins have been identified. Possible groups include sugars, hydroxyl (OH) groups, or ethers (such as OCH_3). These molecules are water-soluble due to the presence of hydroxyl groups and sugars.

PART II ACIDS, BASES, AND pH

Acids are substances that *donate* a hydrogen ion (H^+) in water and form a hydronium ion (H_3O^+). (The H^+, a proton, is often written for convenience, but actually does not exist – the hydronium ion is what we mean when we write H^+ in this context.) A strong acid, for example, hydrochloric acid,

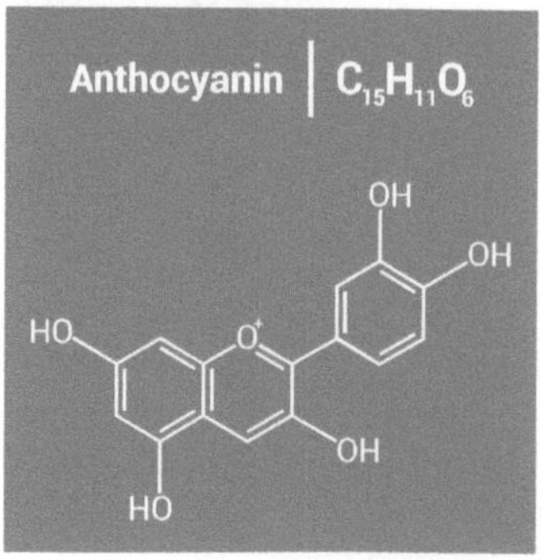

Figure 22.2 The structure of anthocyanin is complex.

releases the H^+ and ionizes completely, or 100%. A weak acid, for example, acetic acid (the salient component in vinegar), does not ionize completely. Only about 5% of the acetic acid molecules give up their H^+ to form the hydronium ion, while 95% of the H^+ remain bonded to the acetate ion.

An important safety concern is that laboratory workers recognize the difference between a strong acid and a concentrated acid. A strong acid may not pose a serious hazard if sufficiently dilute. Similarly, a weak acid may be very hazardous if in concentrated form. It is the total concentration of hydronium ions here that contributes to the hazard of the acid. Additionally, some weak acids such as hydrofluoric (HF) are inherently dangerous and require special handling.

Bases are substances that *accept* an H^+. A good example is ammonia or NH_3. NH_3 will, in water, form the ammonium ion, NH_4^+, and the OH^- group. Many bases in living organisms do not contain the OH^- group. Safe use of bases is similar to that of acids; it is the concentration of base that is the chief concern.

The pH of a solution is a measure of the concentration of hydronium or hydroxyl group present. The chemical definition of pH is:

$$pH \equiv -\log [H^+]$$

The meaning of this equation is that with a decrease in pH by 1 unit, the hydronium ion concentration increases by a factor of 10. (It becomes 10 times more concentrated.) At pH 7, the hydronium and hydroxyl concentrations are equal, but at pH 6, the hydronium concentration is *ten times* greater than the hydroxide ion concentration. The trend continues; at pH 5, the hydronium concentration is one hundred times greater than the hydroxyide ion concentration.

One of the amazing properties of anthocyanins is that this class of plant pigments can change color with pH! The color change is so reliable that it is possible to estimate pH based on the color we see. Plants control their color by the pH of the vacuole, where the anthocyanin is stored. We use this property in today's lab to evaluate pH.

INSTRUCTIONS

Safety. Put on your PPE.

Supplies for 24 students or teams of students. 1 lb red cabbage, preferably shredded, 6 balances to 0.1 g or better; 6 250 mL borosilicate beakers; 12–24 hot plates (preferably) or Bunsen burners; 24 graduated cylinders 50 mL or larger; 336 test tubes; labels or pencils to label test tubes; 50 mL 1.0 M aqueous KCl solution in dropper bottle; deionized or distilled water; 100 mL each of standard solutions of pH 1, 3, 5, 7, 9, 11, and 13 (CAUTION: MOST ARE CORROSIVE); 24 stirring rods; 12–24 heat-resistant gloves or beaker tongs; and various household chemicals (recommend colorless where possible) such as detergent, fruit juice, sodas, milk, plant fertilizer, drain cleaner (CAUTION: CORROSIVE), shampoo, vinegar, tap water.

PART 1 EXTRACTION OF ANTHOCYANIN FROM RED CABBAGE

1. Weigh out 12 grams of red cabbage and if not already shredded, tear into 1 cm^2 pieces.
2. Place the cabbage in a 250 mL beaker and add enough deionized water (about 150 mL) to cover the cabbage.
3. Heat the mixture on a hot plate until the water begins to boil. After 5-7 minutes of boiling time, decant the solution. *Decant* means to pour off the liquid, leaving the solid material behind. It is a faster but somewhat less accurate method of separation than filtration. **CAUTION: USE HEAT-RESISTANT GLOVES OR BEAKER TONGS TO AVOID BURNS.**

You should have about 50 mL of liquid; if not, add enough deionized water to bring the volume up to 50 mL

Note that the extract does not smell as good as fruit juice. Anthocyanin is also present in blueberries; however, red cabbage is much less expensive than blueberries. ☺

PART II DETERMINATION OF THE COLOR OF RED CABBAGE EXTRACT AT SEVEN DIFFERENT PH LEVELS

1. Label test tubes 1, 3, 5, 7, 9, 11, and 13.
2. To each test tube, add about 2 mL of red cabbage juice.
3. Add 3 drops of 1.0 M KCl (potassium chloride) to each tube and mix.
4. **CAUTION: CONSIDER ALL STANDARD SOLUTIONS CORROSIVE (EXCEPT THE pH 7 SOLUTION).** To test tube 1, add 2 mL of pH 1 standard solution. To Test Tube 3, add 2 mL of pH 3 standard solution, and continue until 2 mL of a standard solution has been added to all 7 test tubes. You may mix test tube contents by *gently* tapping the outside of the test tube. Record the colors of your set of test tubes in Table 22.1. Save the test tubes and their contents. They are needed for Part 3.

PART III DETERMINATION OF THE PH OF SEVEN DIFFERENT HOUSEHOLD CHEMICALS

CAUTION: HOUSEHOLD CHEMICALS CAN BE HAZARDOUS. Be sure to rinse your stirring rod thoroughly with deionized or distilled water between samples, and dry with a paper towel.

1. Set up seven more test tubes and label them with the names of the various household chemicals to be tested.
2. Add 2 mL of your cabbage extract and 3 drops of KCl solution to each test tube; mix.
3. Add 2-4 drops of the household chemicals to the appropriately labelled test tubes.
4. Mix and determine the pH of the household chemicals by comparing them with the pH standards. For example, if a chemical's color best matches that of the pH 5 standard solution, the household chemical's pH will be assigned as 5.
5. Record the color of each test tube and the pH assigned in Table 22.2.

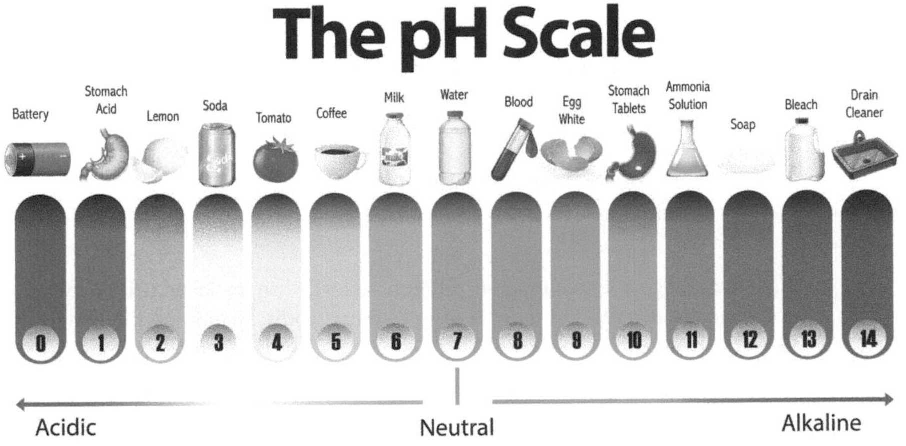

Figure 22.3 The pH scale is only neutral at 7. Excessive acidity and alkalinity can threaten life.

EXERCISE 22 WHAT'S IN RED CABBAGE? ADVENTURES IN PH LAND

Pre-Lab

Last Name ______________________ **First Name** ________________

Instructor ______________________ **Date** ____________________

1. State the chemical definition of pH.

2. State the usual range of pH. ____________________________

3. True or False: All household products have a neutral pH.

4. True or False: All foods have an acidic pH value.

5. State the change in concentration (dilution factor) when an acidic pH value is changed from a pH of 3 to a pH of 4. In other words, what is the dilution factor?
   a. 2 times
   b. 5 times
   c. 10 times
   d. None of these

EXERCISE 22 WHAT'S IN RED CABBAGE? ADVENTURES IN PH LAND

Data

Last Name ______________________ **First Name** ________________
Instructor ______________________ **Date** ____________________

Table 22.1 **Color of Anthocyanin in Various pH Standard Solutions**

pH of Standard Solution	Color of Anthocyanin (red cabbage extract)
1	
3	
5	
7	
9	
11	
13	

1. Comment on the colors of your pH standards (solutions with pH of 1, 3, 5, 7, 9, 11, and 13). Are there any trends in pH and color?

    ____

2. Identify the pH standard you studied which is not corrosive

    ____

Table 22.2 **Common Household Chemicals and their pH Values**

Household Chemical	pH

3. List the household products you studied which have a pH close to neutral.

____

4. List the household products you studied which are most acidic.

____

5. List the household products you studied which are most basic.

____

6. Identify the household chemical you studied which has the highest concentration of H_3O^+.

____

7. Identify the household chemical you studied which has the lowest concentration of H_3O^+.

____

8. What product surprised you most with regard to its pH value?

____

EXERCISE 23

The Process of Extraction – Measuring Caffeine in Tea Bags

Extraction is a useful method to remove harmful as well as beneficial compounds from natural products.

Figure 23.1 An example of the extraction process is the removal of caffeine from tea.

INTRODUCTION

Beverages such as tea (Figure 23.1) can be decaffeinated by the process of **extraction**, with the caffeine then added to sodas. Extraction is also useful for the separation and removal of environmental toxins. As an example, pesticides have

DOI: 10.1201/9781003006565-26

been extracted from beeswax as shown in Figure 23.2 using hexanes and subsequent treatment with an organic chemical called N,N-dimethylformamide (DMF). Although caffeine is not an environmental toxin, this lab does illustrate the process of extraction which generally utilizes either polar or nonpolar solvents. **Polar solvents** have a charge separation, i.e., distinct concentrations of positive and negative charge. In the case of the carbon monoxide molecule, it is polar because the carbon tends to have a very slight positive charge while the oxygen has a very slight negative charge. In contrast, a molecule like hydrogen or H_2 is **nonpolar** because neither hydrogen has a permanent excess of positive or negative charge. Using the principle of "Like dissolves like," we could predict that

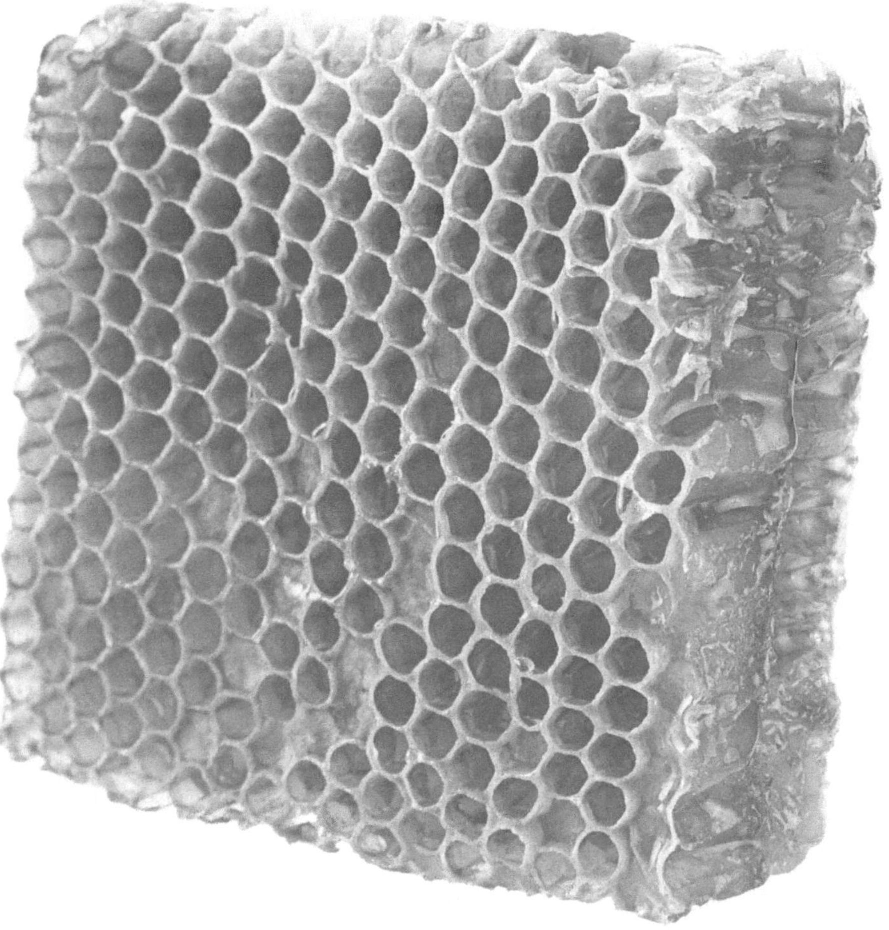

Figure 23.2 Beeswax can be cleaned by an extraction method similar to that used in this lab experiment.

a nonpolar solvent should be used to extract a nonpolar substance. Similarly, to extract a polar substance, a polar solvent such as water should be used. Caffeine, or 1,3,7-trimethylxanthine per its chemical nomenclature, is a Central Nervous System (CNS) stimulant which is consumed by 85% of all Americans – daily. In this lab, we are to extract caffeine from tea bags with aqueous sodium carbonate, followed by the nonpolar solvent methylene chloride (CH_2Cl_2). After the caffeine is extracted, we are to purify it by a process known as **sublimation**. Sublimation is the proess in which a substance changes directly from the solid phase to the gas phase, as follows, Solid ⟶ Gas, without becoming a liquid in between. Carbon dioxide (aka "dry ice") and iodine crystals also sublime at atmospheric pressure.

INSTRUCTIONS

Safety. Put on your PPE.

Supplies for 24 students or teams of students. 24 caffeinated tea bags, 6 centigram or better balances, weighing boats, 120 g anhydrous sodium carbonate, 400 mL methylene chloride, 24 400 mL beakers, 24 100 mL beakers, 24 watch glasses, 24 50 mL or 100 mL graduated cylinders, 24 hot plates, distilled or deionized water, 24 glass stirring rods, 2–4 fume hoods, 12–24 separatory funnels, 100 mL chipped ice, 24 pairs of neoprene or rubber gloves, 12-24 heat resistant gloves, tongs if desired.

PROCEDURE

Weigh a 100 mL beaker, record the mass on line 2 of your Data page, and set it aside for later in the experiment. Take one commercial tea bag and remove any string and/or tag attached to it; weigh the tea bag. It should weigh between 2.1 and 2.9 g. (If the weight is less than 2 g or more than 3 g, consult your instructor as to the amounts of chemicals to use in the rest of the experiment.) Weigh out 3.3 g of anhydrous sodium carbonate (Na_2CO_3) in a weighing boat and add the compound to a 400 mL beaker. Measure 30 mL of deionized/distilled water in a graduated cylinder and add it to the beaker. Protect your hand with a heat resistant glove; then heat the mixture gently on a hot plate in the hood, swirling occasionally while holding the beaker by the rim or upper part, until the solid has dissolved. Add the tea bag to the beaker and use a stirring rod to place the tea bag flat on the bottom of the beaker covered by the sodium carbonate solution. Heat the beaker and contents until the solution begins to boil gently. Allow the solution to boil for 20 minutes.

Allow the beaker and contents to cool to room temperature. (After the beaker has cooled somewhat, you may use a tap water bath to speed up the cooling process.) Remove the tea bag from the beaker, pressing the tea bag gently against the side of the flask to remove the liquid from the bag. Be careful not to handle the tea bag too roughly, as the bag will break and make the procedure more difficult. (The solution is basic and will irritate your skin if it makes contact; neoprene or rubber gloves may be helpful in handling the tea bag and getting as much of the solution out of it as possible.) Transfer the liquid to a separatory funnel and add 6 mL of methylene chloride (dichloromethane, CH_2Cl_2) to the separatory funnel.

You should see two distinct layers (phases) in the funnel. The methylene chloride should be the lower layer, and the aqueous sodium carbonate solution should be on top. Stopper the funnel and shake it *gently* to mix the two liquids. This will caus the caffeine which is nonpolar, to move from the aqueous layer which is polar, to the nonpolar methylene chloride. (Your instructor may demonstrate this technique.) Be sure to vent the funnel occasionally by removing the stopper to release the pressure that builds up. Protect your eyes! Do not look in the funnel through the upper opening while it is unstoppered. Remove the stopper and allow the two layers to separate. Drain the methylene chloride lower layer through the stopcock into the 100 mL beaker that you already weighed. Add 6 mL of fresh methylene chloride to the separatory funnel and repeat the shaking process described above. Extraction processes often require multiple steps. We are using two steps here. After the layers settle again, drain the methylene chloride into the beaker containing the first 6 mL of methylene chloride.

Place the beaker containing the methylene chloride solution on a hot plate in the hood and *gently* evaporate the methylene chloride solvent away from the caffeine. Do not heat the beaker too strongly, or the caffeine will be lost. [This is an all too common error.] Stop the evaporation process while the caffeine still still looks damp; residual heat will dry it. When all of the solvent has been removed, you should see your crude (unpurified) caffeine as a white or yellowish solid in the bottom of the beaker. Using a heat resistant glove or tongs, remove the beaker from the hot plate and allow the beaker to cool to room temperature. Weigh the beaker, record the value on line 1 of your data page, and subtract the weight of the empty beaker to determine the weight of your crude caffeine. Record this weight of the crude caffeine on line 3 of your data page. Describe the appearance of the caffeine on line 4.

To purify your caffeine by sublimation, place your beaker of caffeine on a hot plate and cover the beaker with a watch glass with the concave side of the watch glass up. Place a small piece of ice on the watch glass. Very gently heat the beaker and look for a solid to form on the underside of the watch glass. Describe what you saw on line 5 of your data sheet.

EXERCISE 23 THE PROCESS OF EXTRACTION – MEASURING CAFFEINE IN TEA BAGS

Pre-Lab Questions

Last Name ______________________ **First Name** ________________
Instructor ______________________ **Date** ______________________

1. Name the chemical used to extract caffeine from tea.

2. After the caffeine is extracted from the tea, what is the organic chemical used to remove the caffeine from the extraction solution?

3. Why should gloves be used during this experiment?

4. Explain why part of this experiment is to be conducted under a fume hood.

5. True or False: All brands of tea contain the same amount of caffeine.

6. True or False: As pesticides have successfully been extracted from beeswax, the process could also logically be applied to extracting pesticides from sheep wool wax (lanolin).

7. What do the processes indicated in Question 6 tell us about the ubiquity of pesticides in our environment?

EXERCISE 23 THE PROCESS OF EXTRACTION – MEASURING CAFFEINE IN TEA BAGS

Data

Last Name ______________________ **First Name** _________________
Instructor _______________________ **Date** _____________________

1. **Weight of (beaker + caffeine):** _________ **g**
2. **Weight of empty beaker:** _________ **g**
3. **Weight of crude caffeine:** _________ **g**
4. **Description of crude caffeine:**

   ____

5. **Observations on sublimation:**

   ____

   ____

EXERCISE 23 THE PROCESS OF EXTRACTION – MEASURING CAFFEINE IN TEA BAGS

Post-Lab Questions

Last Name ______________________ **First Name** ______________

Instructor ______________________ **Date** ______________________

1. True or False: All teas have the same concentration of caffeine.

2. What is the name of the type of funnel used to separate the extracted caffeine from the hot water?

3. What is the function of methylene chloride in this experiment?

4. What does the term sublimation mean?

5. What is the expected color of the crude caffeine which you extracted?

6. Pesticides are generally nonpolar. Which solvent would be most effective in removing pesticides by extraction from a natural product such as lanolin?
   a. Hexanes (nonpolar)
   b. Water (polar)
   c. Ethanol (slightly polar)
   d. Acetone (slightly polar)

BIBLIOGRAPHY

Calatayud-Vernich, Pau, VanEngelsdorp, Dennis, and Pico, Yolanda, Beeswax cleaning by solvent extraction of pesticides, *MethodsX*, *6*, 980–985. doi:10.1016/j.mex.2019.04.022

Mitchell, Diance C., et al., Beverage caffeine intakes in the U.S. *Food and Chemical Toxicology*, 2014, *63*, 136–142. doi:10.1016/j.fct.2013.10.042

EXERCISE 24

Soap or Detergent: Which Is Better? Preparing Soap

The increased use of soaps during the COVID-19 pandemic led to environmental problems.

Figure 24.1 Soap is not a new idea. People have been making soap from fats for millennia.

INTRODUCTION

Although soaps date back to ancient Babylonian times, and their popularity is well established, their increased use during the COVID-19 pandemic led to environmental problems. Yes, people were washing their hands more often, and for good reasons. This of course increased the use of soap. While preventing viral infection was a priority, the increased packaging waste and detergent wastewater presented environmental challenges. This illustrates the Law of Unintended Consequences which states that actions taken to solve *one* problem can often create a *new* problem.

High detergent concentrations in freshwater can increase foam which reduces the ability of oxygen to dissolve in the water, changes pH and other variables, and ultimately threatens marine animals. Aquatic plants are also endangered by high concentrations of detergents in freshwater, through **eutrophication** which is

DOI: 10.1201/9781003006565-27

excessive growth of algae, etc. at the expense of other more desirable species. Antibacterial additives such as triclocarban (TCC) and triclosan (TCS) to soaps can lead to strains of bacteria which resist antibiotics.

So, exactly what are detergents and what are soaps, and what is the difference? A detergent can be defined as anything that helps water do a better job of cleaning. Soap, as shown in Figure 24.1, is a detergent, and is made from fats or oils combined with strong base (alkali).

Synthetic detergents tend to be made from petroleum and other ingredients. They are salts of benzenesulfonic acid. Sometimes they contain brighteners which absorb and reflect light, making the clothes look cleaner than they are. Detergent molecules, soap included, have a **hydrophilic** (polar) portion that attracts water and a **hydrophobic** (nonpolar) portion that helps dissolve grease, etc.

Soaps have been a common household item for many hundreds of years. While they were well known throughout the Mediterranean area during the Middle Ages, they were not made commercially until the nineteenth century. Surplus fats and oils were boiled in a water solution of a basic substance (alkali), which was usually obtained from the ashes of wood or seaweed. The average rural American family of the early twentieth century made most of their soaps. The fats were obtained from the kitchen. Lye, a commercial grade of sodium hydroxide (NaOH), was used as the alkali. During boiling, the fats and oils, which are esters of glycerol and long-chain fatty acids, undergo hydrolysis to form glycerol and soap. Soaps are alkali metal salts of long-chain fatty acids. Hydrolysis of esters in an alkaline solution is called saponification (literally, "soapmaking").

Grease consists of long-chain hydrocarbons which are relatively nonpolar; dirt is also nonpolar. Water is a polar substance and will not dissolve "greasy dirt." To dissolve grease and dirt, we need to add something to water that is both nonpolar and polar. Anions of fatty acids (soap) have both a long nonpolar tail and a polar head.

Ions such as these can be dispersed in water because they form micelles as shown in Figure 24.2. The nonpolar tails associate in the interior of the micelle to break up oil or grease, while the polar heads point outward to interact with and remain dissolved in the polar water molecules. A soap "solution" is therefore not a true solution but is a dispersion of these micelles in water, making the water look cloudy.

A major disadvantage of soap is that it forms precipitates with the cations found in hard water, principally Ca^{2+} and Mg^{2+}. This "soap scum" is responsible for "ring around the collar," "bathtub rings," and more. By tying up the soap ions, hard water ions decrease soap's cleaning efficiency. The development of synthetic detergents, using petroleum to form the long carbon chains, brought about a new product that would clean well even in hard water. The anions of synthetic detergents have the advantage of not forming insoluble precipitates with calcium and magnesium ions. Figure 24.3 shows the concentrations of hardness of water throughout the United States. Clearly, areas vary in this respect throughout the country.

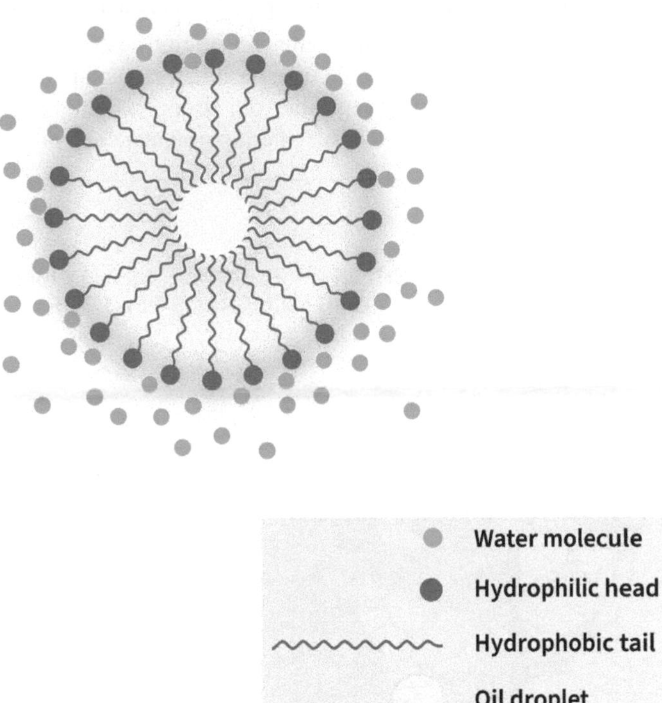

Figure 24.2 Action of a detergent molecule to form a micelle.

INSTRUCTIONS

Safety.

* Put on your PPE.
* The 6 M NaOH is caustic, so wash your hands well if they come into contact with it. Bases are slippery on the skin.
* Ethyl alcohol is highly flammable.
* Your soap will have an alkaline residue. Avoid contact, especially with the eyes.
* Dispose of any waste solutions as directed by your instructor.

Supplies for 24 students or teams. 6 balances, 0.1 g or better, weighing boats, 500 mL ethanol (95%), 250 mL of 6 M sodium hydroxide solution CAUTION: Corrosive, 12–24 hot plates, shortening or oil or both for comparison of products, sodium chloride, ice, 12 50 mL or 100 mL graduated cylinders, commercial soap, 24 glass stirring rods, 24 small test tubes, 24 250 mL beakers, 24 600 mL beakers, paper towels,* deionized/distilled water, conical funnel,* ringstand,* filter paper*, 48 small test tubes, extra beaker or test tube rack to hold test tubes, 24 50 mL beakers, 25 mL phenolphthalein indicator solution, dropper bottle recommended, 50 mL commercial liquid household (not lab) detergent, 25 mL "hard water," e.g. 1 M magnesium or calcium nitrate, dropper bottle recommended, 25 mL vegetable oil, dropper bottle recommended 12 heat resistant gloves. *If needed for filtering granular soap product.

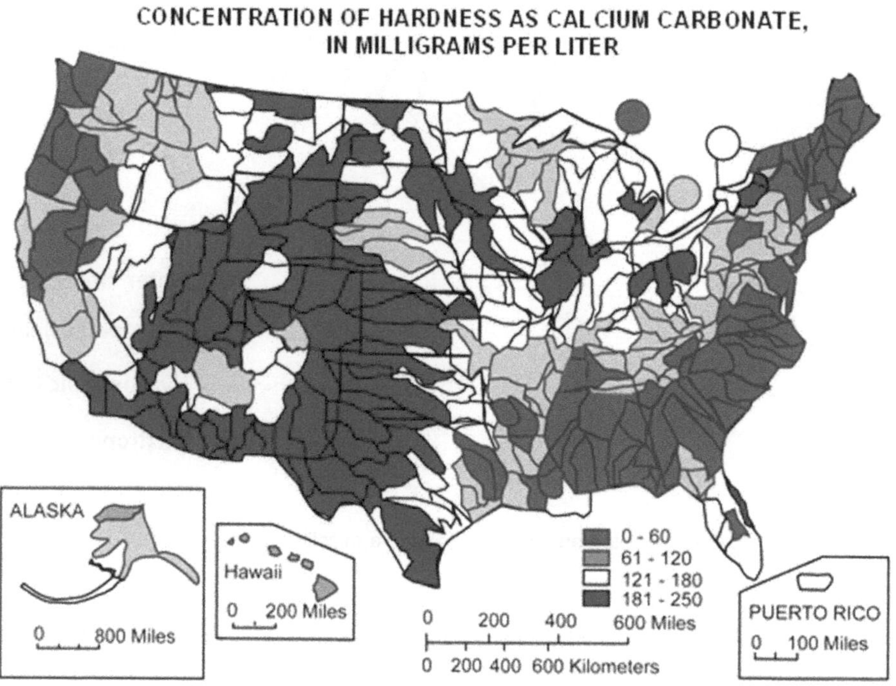

Figure 24.3 Map of water hardness in the United States. www.usgs.gov/media/images/map-water-hardness-united-states

INSTRUCTIONS

Part I. Preparation of Soap. Weigh 9 g of sodium chloride on the balance in a weighing boat. Measure 25 mL of deionized/distilled water into a 50 mL beaker. Place the beaker on a hot plate and heat the water without bringing it to a boil. Slowly add the 9 g of sodium chloride, stirring with a glass rod. Then add about 300 mL of tap water to your 600 mL beaker and bring to a boil on the hot plate. Next, weigh out 5 g of the fat or oil provided in a 250 mL beaker and add 8 mL of 6 M NaOH **CAUTION: CORROSIVE** and 20 mL ethanol **CAUTION: FLAMMABLE**. Stir well and place the beaker and its contents carefully in the 600 mL beaker. Lower the hot plate setting so that the beaker has barely enough heat to boil. Stir the mixture from time to time so that the contents of the small beaker do not remain separated. If a significant volume is lost, add deionized/distilled water to bring the mixture back up to its original volume, but no more. After 30 minutes remove a little bit of the material with a stirring rod and dissolve it in a test tube of warm water by shaking it vigorously. If good suds are produced and there is no evidence of free fat remaining, then add 10 mL of deionized/distilled water and then the hot saturated salt solution you prepared earlier, to the 250 mL beaker of soap. The soap will separate out into a **homogenous** (same composition throughout) layer. Let the soap cool and remove it from the beaker after it has become solid. If a granular product results from the addition of the salt solution, then filter it and press the granules between layers of paper towels.

Part II. Soap and Commercial Detergent Reactions. Testing Soap with Phenolphthalein. Warm about 20 mL of deionized/distilled water in a 50 mL beaker on a hot plate. Place a test tube in an empty beaker or test tube rack and use a heat resistant glove to pour the warm water in the test tube. Add a little of the soap you prepared to the test tube, stirring with a glass rod to dissolve it. Add one drop of phenolphthalein indicator solution to the test tube. This indicator is colorless in acid solution but fuschia pink in base solution. Was there a color change? What do the results mean? Record your observation and conclusion as to whether soap is acidic or basic on your data sheet, line 1.

Testing Commecial Detergent with Phenolphthalein. Repeat the process above, but use commercial liquid detergent. Was there a color change? What do the results mean? Record your observations and conclusions on your data sheet, line 2.

Testing with Hard Water. Place two clean test tubes in a test tube rack or beaker. Fill one test tube about 1/2 full with deionized/distilled water, and dissolve a little of your prepared soap in the water. Repeat with a second test tube, using a little liquid detergent instead of the soap you prepared. Add a few drops of hard water to each test tube. What happened in each test tube? Record the results on your data sheet, line 3.

Testing for Interaction with Oil. Place a small test tube in your test tube rack or a beaker. Fill it about one-half full of the warm deionized/distilled water you prepared above, and add two drops of vegetable oil Shake the tube vigorously and let sit for 10 minutes. Did the oil mix with the water, or form a layer on top? Record your observations on your data sheet, line 4.

Dispose of any left over soap as directed by your instructor. Avoid eye and skin contact with the soap as it may be corrosive due to excessive alkalinity.

EXERCISE 24 SOAP OR DETERGENT: WHICH IS BETTER? PREPARING SOAP

Pre-Lab Questions

Last Name ______________________ **First Name** __________________
Instructor ______________________ **Date** ______________________

1. True or false? Soap is a relatively new invention.

2. Define saponification.

3. List the two main ingredients needed to prepare soap.

4. List the two ions commonly responsible for hard water.

5. Which work well in hard water?
   a. Soaps
   b. Synthetic detergents
   c. Both soaps and synthetic detergents
   d. None of these
6. Excess soap and synthetic detergents in the environment can be hazardous for
   a. aquatic plants
   b. marine animals
   c. aquatic plants and marine animals
   d. none of these

EXERCISE 24 SOAP OR DETERGENT: WHICH IS BETTER? PREPARING SOAP

Data

Last Name ______________________ **First Name** __________________
Instructor ______________________ **Date** ______________________

1.

2.

3.

4.

EXERCISE 24 SOAP OR DETERGENT: WHICH IS BETTER? SYNTHESIZING SOAP

Post-Lab Questions

Last Name ____________________ **First Name** ____________________
Instructor ____________________ **Date** ____________________

1. What was the purpose of adding the salt solution to the initial soap reaction mixture?

2. Explain how polar and nonpolar molecules differ.

3. Sketch a soap molecule and label the polar and nonpolar parts.

4. Fill in the blanks: The ____________________ part of a soap molecule will mix with grease and the ____________________ part will mix with water.

5. Explain the role of soaps and other detergents in removing grease from dishes and clothing.

6. Does the expression "like dissolves like" apply to the action of soaps and other detergents? Explain.

7. What kind of solution forms when soap dissolves in water?

8. What is present in soapy water that causes its cloudy appearance?

9. List the chemical components of hard water.

10. Identify the source of bathtub rings and "ring around the collar."

11. Which functions better in hard water, soap or a synthetic detergent? Why?

BIBLIOGRAPHY

Bruice, Paula Yurkanis, *Organic Chemistry*, Eighth Edition, Pearson Education, Upper Saddle River, NJ, 2016, p. 1134.

Buell, Phyllis and Girard, James, *Chemistry: An Environmental Perspective*, Prentice-Hall, Inc., Englewood Cliffs, NJ, 1994, p. 302.

Chirani, Mahboobeh Rafieepoor, et al., Environmental impact of increased soap consumption during COVID-19 pandemic: biodegradable soap production and sustainable packaging, *Science of the Total Environment*, 2021, *796*, 1–11. doi:10.1016/j.scitotenv.2021.149013

USGS, Map of water hardness in the United States, www.usgs.gov/edia/images/map-water-hardness-united-states

EXERCISE 25

What's in My Detergent? Measuring Phosphate with Spectroscopy

With too much phosphate ion, algae and excessive weed growth can take over natural waters.

Figure 25.1 Laundry detergents offer advantages over soap, and some are formulated to be environmentally friendly.

INTRODUCTION

Over the years detergents like those shown in Figure 25.1, have been very popular, but somehave added significantly to the concentrations of phosphate (PO_4^{3-}) ions in natural waters. This is due to the former use of phosphate-containing compounds in detergents as "builders" (compounds that help the detergent work better); however, currently, phosphates are not used for this purpose nearly as much as they were before the environmental problem of excess phosphate was noticed. Rather than being a toxin to plants and animals, phosphate helps plants grow! Phosphates are also found in fertilizers. With too much phosphate, algae and excessive weed growth can take over natural waters, crowding out other more desirable species. **Eutrophication** is the technical term

DOI: 10.1201/9781003006565-28

for this problem of plant overgrowth encouraged by excess phosphate. When the plants outstrip their food supply, they die and create a high biological oxygen demand (BOD) and drastically lower the level of dissolved oxygen (DO) in the water, causing other aquatic life (both plants and animals) to die. In this lab we can analyze these solutions by the extent to which they absorb red light of wavelength 650 nanometers (nm or 10^{-9} m). [Our eyes can detect light of wavelengths 400 (violet)–700 (red) nm.] First, we will check solutions of *known* concentrations – these are called **standards** – and they will allow us to determine the concentrations of *unknown* solutions. Just like we use a ruler or meter stick of known length, to determine unknown lengths, we use standard solutions of known concentrations to determine the concentrations of unknown solutions.

INSTRUCTIONS

Safety. Put on your PPE.

Supplies for 24 students or teams. 6 visible-range spectrophotometers, cuvettes (recommend disposable), 1 L of stock phosphate solution (20.0 ppm), ammonium molybdate reagent, 200 mL stannous chloride reagents, up to 300 mL each of samples of unknown phosphate concentration, 1 mL Beral pipettes, 24 10 mL graduated cylinders, 24 50 mL or 100 mL graduated cylinders, 24 glass stirring rods, 175 125 mL or 250 mL Erlenmeyer flasks, deionized/distilled water. Students may work in teams of four each, reducing the quantity of supplies needed.

1. *Preparation of Standards*. Prepare the following standard solutions according to Table 25.1. For each solution, measure the amount of phosphate stock solution shown in Table 25.1, in a 10 mL graduated cylinder. Transfer to a 50 or 100 mL graduated cylinder and add sufficient deionized water to bring the final volume up to exactly 40 mL. Measure volumes carefully, and mix well with a glass stirring rod, to get good results!

Table 25.1 Preparation of phosphate standard solutions

Standard	Volume of 20 ppm Phosphate Stock Solution (mL)	Final Volume (mL)	Phosphate Concentration (ppm)
A	2.0	40.0	1.0
B	4.0	40.0	2.0
C	6.0	40.0	3.0
D	8.0	40.0	4.0
E	10.0	40.0	5.0

2. *Preparation of Solutions to Analyze.* Prepare the following solutions in Erlenmeyer flasks, according to Table 25.2. Again, measure and mix carefully so you will get good results. Phosphate ion, if present, will cause a blue: color to appear and reach its maximum intensity in 5 minutes or less. These must be analyzed in the spectrophotometer between 5 and 15 minutes after stannous chloride (CAUTION: TOXIC) is added. Do not add the stannous chloride solution until you are sure there is a spectrophotometer available so you can measure the absorbance within 15 minutes of the mixing process.

Table 25.2 Preparation and absorbance readings of phosphate solutions

To Be Analyzed	Ammonium Molybdate Solution (mL)	Stannous Chloride Solution	Absorbance
Phosphate Standard A, 1 ppm, 25 mL	1	2 drops	________
Phosphate Standard B, 2 ppm, 25 mL	1	2 drops	________
Phosphate Standard C, 3 ppm, 25 mL	1	2 drops	________
Phosphate Standard D, 4 ppm, 25 mL	1	2 drops	________
Phosphate Standard E, 5 ppm, 25 mL	1	2 drops	________
Blank, 25 mL deionized water (no phosphate)	1	2 drops	Zero
Unknown 1, 25 mL	1	2 drops	________
Unknown 2, 25 mL	1	2 drops	________

3. *Collection of Data.* Be sure that your spectrophotometer is warmed up and set to a wavelength of 650 nm. Add some of the blank solution to a clean cuvet, until the liquid level is about 2 cm from the top. Dry the exterior of the cuvet with a Kim-Wipe™. These do not leave lint and do not scratch the delicate optical glass – do not use a paper towel. Open the spectrophotometer's sample compartment, place the cuvet in the spectrophotometer (without spilling contents), gently close the sample compartment, and set the absorbance to Zero. Next, place each standard in a clean cuvet, and measure and record their absorbance in Table 25.2. Cuvets must be clean and dry, or clean and rinsed with sample solution before they are filled. Finally, determine and record the absorbances of the unknowns in the same way.
4. *Analysis of Data.* Make a graph of the absorbance of each standard sample (on the ordinate which is the y or vertical axis) vs the concentration of each standard sample (on the abcissa which is the x or horizontal axis). Draw the best possible straight line through the points. Next, take the absorbance of the first unknown solution, locating it on the straight line, and going down to the x (phosphate concentration) axis to determine the concentration. Repeat for the second unknown. Record the concentrations of the unknown solutions in Table 25.3.

EXERCISE 25 WHAT'S IN MY DETERGENT? MEASURING PHOSPHATE WITH SPECTROSCOPY

Pre-Lab Questions

Last Name ____________________ **First Name** ____________________

Instructor ____________________ **Date** ____________________

1. Name the process which refers to overgrowth of algae and other marine plants which threaten more desirable species in our natural waters.

2. Name the ion which contributes strongly to the process in question one.

3. True or False? A high biological oxygen demand (BOD) helps protect aquatic life.

4. Today's lab analyzes the ability of phosphate solutions to absorb light of wavelength _________ nm.

5. In this experiment solutions of known concentrations are called

EXERCISE 25 WHAT'S IN MY DETERGENT? MEASURING PHOSPHATE WITH SPECTROSCOPY

Data

Last Name ______________________ **First Name** __________________

Instructor ______________________ **Date** ________________________

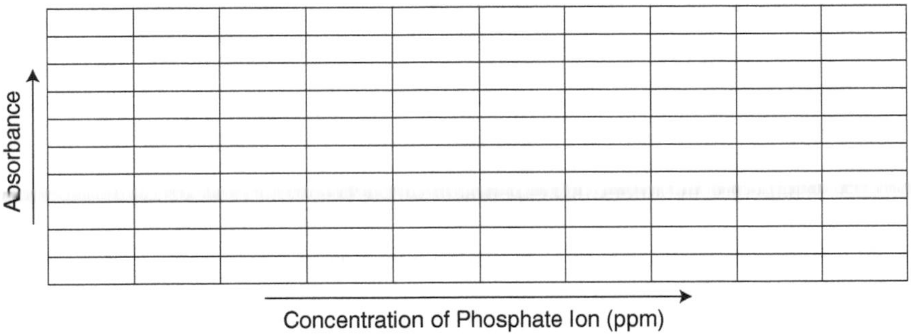

Absorbance

Concentration of Phosphate Ion (ppm)

Graph 25.1 Graph of absorbance vs concentration of phosphate ion.

Table 25.3 Phosphate concentrations of unknown solutions

Sample	Phosphate Concentration (ppm)
Unknown 1	__________________
Unknown 2	__________________

EXERCISE 25 WHAT'S IN MY DETERGENT? MEASURING PHOSPHATE WITH SPECTROSCOPY

Post-Lab Questions

Last Name ____________________ **First Name** __________________
Instructor _____________________ **Date** ______________________

1. Name at least two common household products which contain phosphates.____
2. Explain what happens when excess phosphates are released to rivers and lakes. ____
3. What color light does this experiment utilize?______________
4. The concentration of the unknown sample is determined by
   a. calculation
   b. interpreting values from a graph
5. In this experiment, you must wait until the reagents develop the color needed, before completing the analysis.
   a. True
   b. False

EXERCISE 26

What's in Water Besides H_2O? A Laboratory Analysis

Our bodies are about 70% water.

Figure 26.1 Water is important for life and, as it provides recreation, for quality of life.

INTRODUCTION

Water aka H_2O is super-important! We drink it, bathe in it, and swim in it, making sites like the one shown in Figure 26.1 very popular. Additionally, the molecule water shown in Figure 26.2 makes up about 70% of our bodies! Clearly, the water we use for these activities must be safe! This lab shows us,

DOI: 10.1201/9781003006565-29

on a small scale, *some* of the steps involved in purifying water. It does not involve chlorination to kill bacteria, so even if you do the lab perfectly, do not drink the water. As you probably know by now, consuming anything in or from the lab is not allowed. This exercise involves the analysis of (1) dissolved solids (sometimes called TDS for total dissolved solids) and (2) suspended solids (TSS for total suspended solids, which are those which can be removed by filtration). Your instructor may ask you to bring in water samples from various sources, or may simply provide you with some cloudy water to purify. If you do collect water samples from local streams, etc., watch for snakes and other outdoor hazards!

Your water sample may contain carbonate ions. You can test for this by adding a few drops of hydrochloric acid to the TDS residue, which will release bubbles of carbon dioxide gas according to the following net equation:

$$2H^+ + CO_3^{2-} \rightarrow H_2O + CO_2(g)$$

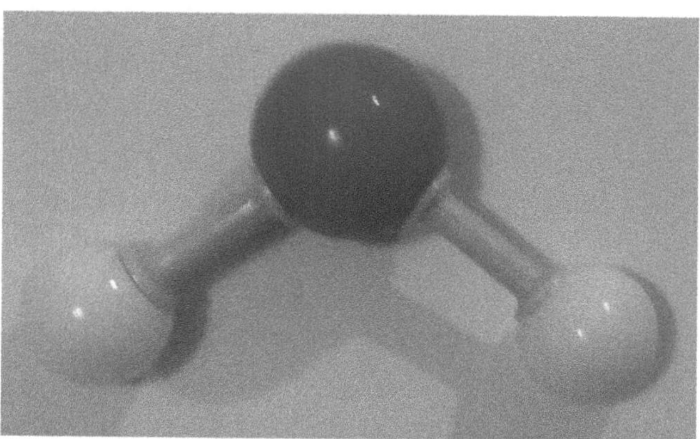

Figure 26.2 The water molecule is essential for life.

INSTRUCTIONS

Safety. Put on your PPE.

Supplies for 24 students or teams. Filter paper, 24 110 mL samples of natural water or equivalent preferably in plastic bottles with cap, deionized/distilled water, 24 150 mL beakers, 24 100 mL graduated cylinders, 24 watch glasses, 6 balances to 0.01 g or better, drying oven at 110 °C (optional), 25 mL of 6 M HCl (CAUTION: CORROSIVE) in dropper bottle, 24 conical funnels, 24 ring stands, 12–24 hot plates (preferred) or Bunsen burners with wire gauze with sintered glass center and matches, 24 heat-resistant gloves or beaker tongs.

Weigh a piece of filter paper and record its mass on line 2 of Table 26.1. Weigh a 150 mL beaker and record the value on line 2 of Table 26.2. Your sample will contain suspended solids and thus not be homogeneous, so it must be mixed to get a representative portion, before some is removed for analysis. Stir (or shake if in a securely stoppered bottle) your water sample, and measure 100 mL of this in a graduated cylinder. Fold the weighed filter paper in half, and then almost in half. Open it so that there are three layers of paper on one side and one layer on the other side; place in the funnel. You may add a few drops of deionized/distilled water to help the filter paper adhere to the funnel. For faster and more effective filtration, the paper should adhere closely to the funnel. Filter your water sample through the weighed filter paper into the weighed beaker. Dry the filter paper in the oven on a watch glass. The oven should be set at about 110 °C to ensure the evaporation of water which boils at 100 °C. Evaporate the water from the beaker on the hot plate. Remove the beaker with a heat-resistant glove or beaker tongs, from the hot plate before it is totally dry; residual heat will complete the drying process. Leaving the beaker on the hot plate until totally dry will cause hazardous splattering and loss of sample. Allow each to cool to room temperature; then re-weigh the filter paper and beaker. Record the mass of the filter paper + residue on line 1 of Table 26.1 and the mass of the beaker + residue on line 1 of Table 26.2. Complete the data page filling in the calculations blanks as shown. Test the residue in the beaker for carbonate ion with a few drops of dilute hydrochloric acid (HCl) (**CAUTION: CORROSIVE**). The presence of carbonate ion is confirmed by the formation of bubbles of carbon dioxide. Record your observations on your data page, #3, "Results of carbonate test ."

EXERCISE 26 WHAT'S IN WATER BESIDES H_2O? A LABORATORY ANALYSIS

Pre-Lab Questions

Last Name ____________________ **First Name** ____________________
Instructor ____________________ **Date** ____________________

1. What two types of solids are tested for in today's lab?

2. What corrosive chemical is used?

3. Why should the beaker be removed from the hot plate before it is totally dry?

4. Why should the water sample be stirred or shaken before the 100 mL to analyze is measured?

5. Why should the filter paper adhere closely to the funnel?

EXERCISE 26 WHAT'S IN WATER BESIDES H_2O? A LABORATORY ANALYSIS

Data

Last Name ____________________ **First Name** ____________________
Instructor ____________________ **Date** ____________________

Table 26.1 **Determination of concentration of suspended solids in water sample**

Suspended Solids	Trial 1	Trial 2
1. Mass of filter paper + residue (g)		
2. Mass of filter paper (g)		
3. Mass of residue (g) (Line 1 - Line 2)		
4. Mass in g/L. Divide g of residue in Line 3 by 0.1 L		
5. Mass in mg/L. Multiply answer in g/L in Line 4 by 1000		

Table 26.2 **Determination of dissolved solids in water sample**

Dissolved Solids	Trial 1	Trial 2
1. Mass of beaker + residue (g)		
2. Mass of beaker (g)		
3. Mass of residue (g) (Line 1 - Line 2)		
4. Mass in g/L. Divide g of residue in Line 3 by 0.1 L		
5. Mass in mg/L. Multiply answer in g/L in Line 4 by 1000.		

1. Source of water sample

2. Date of collection of water sample

3a. What did you see when the HCl was added to the residue in the beaker?

3b. What does this indicate about the presence of carbonate ion in the sample?

4a. Calculate the total solids in the sample.

4b. Calculate the percentage of dissolved solids.

Your Answer. ____________________

5. Calculate the percentage of suspended solids.

Your Answer. ____________________

EXERCISE 26 WHAT'S IN WATER BESIDES H_2O? A LABORATORY ANALYSIS

Post-Lab Questions

Last Name ______________________ **First Name** ________________
Instructor ______________________ **Date** ______________________

Select the *best* answer on multiple-choice questions.

1. Write the balanced equation for the reaction of hydrochloric acid with calcium carbonate.

2. What does the above reaction tell you about the possible action of acid rain on marble and limestone structures, both of which contain carbonate ion?

3. A clear lake is more likely to contain
   a. suspended solids in high concentrations
   b. dissolved solids in high concentrations

4. Why should the drying oven in this experiment be set above 100 °C?

5. In this experiment, dissolved solids would be found in the
   a. residue
   b. filtrate

BIBLIOGRAPHY

Beard, James M., *Environmental Chemistry in Society*, Second Edition, Taylor & Francis, Boca Raton, FL, 2013, p. 268.

EXERCISE 27

Purifying Cloudy Water – A Treatment Process

"In our attempts to dispose of wastes, we often will contaminate ground water."

– James M. Beard, Ph.D. Chemist

Figure 27.1 Beautiful, clear water on the shore of Lake Tahoe on the California-Nevada state line.

INTRODUCTION

Pure water is necessary for esthetics as well as health. Who wants to look at, or smell, polluted water? In contrast to Lake Tahoe in Figure 27.1, the Cuyahoga River in Cleveland, Ohio, was so polluted that it earned the dubious distinction of being called "The River that Burned!"

DOI: 10.1201/9781003006565-30

Water has been called "the solvent of life," and as mentioned in Exercise 26, humans are about 70% water! Truly, water is important! In this lab, we gain experience in *clarifying* water; some bacteria and viruses will be removed in the mechanical action involved; however, we will not add chlorine or a similar treatment which would kill bacteria and possibly viruses. Before we use water for drinking, it must be treated to eliminate the threat of bacteria. Similarly, swimming pools require such treatment. An interesting alternative to chlorination of swimming pools is the saltwater pool. Salt, sodium chloride, in the pool promises to be gentler than chlorine to the eyes and skin. Electrolysis in this application uses electrical energy to convert the chloride ion from the salt, to chlorine.

Please note that after our lab work the water will still not be potable (safe to drink). It is a "rule of the road" to never consume anything in a lab anyway.

PART I: PREPARATION OF ALUMINUM HYDROXIDE, A FLOCCULANT FOR REMOVAL OF SUSPENDED SOLIDS

One of the steps that must be performed on water to make it suitable for drinking is the removal of any suspended solid material from the water. **Suspended solids** are materials such as very fine dirt particles that are carried along by, but not truly dissolved in the water. One way to remove suspended particles is to use a type of chemical called a **flocculant** that combines with the suspended solids to form a mass or floc that is solid enough to be filtered out of the water. Water treatment plants use several different flocculants to remove suspended solids. In this experiment, you will prepare one flocculant, aluminum hydroxide [$Al(OH)_3$] from aluminum metal and then use the aluminum hydroxide to clarify (remove the suspended solids from) some cloudy water. Converting used aluminum cans to this flocculant aluminum hydroxide is one possible use for the old cans, but turning the cans into new aluminum cans is a more economical way to recycle the aluminum. Recycling aluminum rather than preparing more from aluminum ore reduces energy use by about 95%.

Water treatment plants, in addition to flocculation processes, treat water in other ways, as shown in Figure 27.2, before it reaches our homes.

In this experiment we can analyze our water samples with a spectrophotometer set to produce (green) light of 500 nanometers (nm) wavelength. [Visible light ranges from 400 nm (violet) to 700 nm (red).]

The reactions involved in this experiment are as follows:

1. Dissolving aluminum metal with sodium hydroxide solution:

$$2Al + 2Na^{+} + 2OH^{-} + 6H_2O \rightarrow 2Na^{+} + 2Al(OH)_3 + 3H_2(g) \qquad (27.1)$$

2. Converting the aluminum ion into aluminum sulfate (aluminum and sulfate ions in solution):

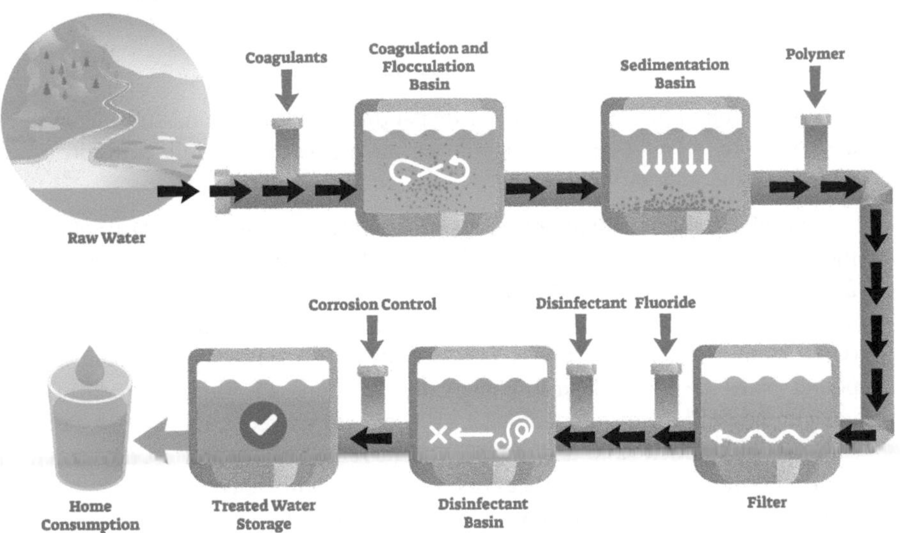

Figure 27.2 Diagram of a water treatment plant. In the Flocculation Basin, the aluminum hydroxide forms "flocs" allowing suspended matter to settle out in the Sedimentation Basin.

$$2Al(OH)_3 + 3H_2SO_4 \rightarrow 2Al^{3+} + 3SO_4^{2-} + 6H_2O \quad (27.2)$$

3. Converting the aluminum sulfate solution into solid aluminum hydroxide (the flocculant):

$$2Al^{3+} + 6OH^- \rightarrow 2Al(OH)_3(s) \quad (27.3)$$

The hydroxide ions used in Step 3 come from the reaction of sodium hydrogen carbonate (also known as sodium bicarbonate or baking soda) when it is dissolved in water:

$$NaHCO_3(s) + H_2O \rightarrow Na^+ + OH^- + CO_2(g) + H_2O \quad (27.4)$$

PART II: TREATMENT WITH ACTIVATED CHARCOAL TO REMOVE DISSOLVED CHEMICALS

Another category of materials that sometimes must be removed from water before it can be used for drinking is **dissolved chemicals**. If the dissolved molecules or ions are large enough, they can be removed by passing the water over a form of carbon called activated charcoal. Activated charcoal has a very large surface area and this surface is able to latch onto (adsorb) certain dissolved materials and hold onto them. The charcoal, along with the chemicals it has caught, can then be filtered out of the water. After the charcoal has been filtered from the water, it can be recycled by heating it with very hot steam (500–1000°C). This causes the charcoal to release the adsorbed chemicals and leaves clean charcoal that can be used again.

INSTRUCTIONS

Safety. Put on your PPE. Remember that sodium hydroxide and sulfuric acid are corrosive. Avoid eye and skin contact, and rinse immediately with lots of water if some gets on you.

Supplies for 24 students or teams. 6–12 balances centigram, 24 Buchner funnels, filter papers to fit funnels, 24 25 or 50 mL graduated cylinders, 24 250 mL beakers, 24 150 mL beakers, 24 100 mL beakers, 24 glass stirring rods, 10 g aluminum foil, 12 pairs of scissors, 500 mL 6 M NaOH (CAUTION; Corrosive!), 600 mL 3 M H_2SO_4 (CAUTION: Corrosive!), 1 pound $NaHCO_3$, 50 mL methylene blue solution, 24 filtering flasks, 24 vacuum hoses, 12 water aspirators, 15 g activated charcoal, 24 "dirty water" samples with added sand and/or dirt, vial of red litmus paper, 4 centrifuges with centrifuge tubes, 4 spectrophotometers with cuvets (recommend disposable), 48 Erlenmeyer flasks, 6 hot plates, 3 fume hoods, deionized/distilled water.

PROCEDURE

I. Preparation and Use of Aluminum Hydroxide

1. Weigh 0.20 g of aluminum foil and place it in a 250 mL beaker.
2. Measure 15 mL of 6 M sodium hydroxide (NaOH) **(CAUTION: Corrosive!)** solution in a graduated cylinder, and pour it over the aluminum foil in the beaker. The foil should be completely covered by the solution.
3. Put the beaker on a hot plate in the fume hood and *gently* heat the mixture. Do not boil. Stir the mixture with a glass stirring rod to keep it from foaming over.
4. Continue warming the solution until most of the fizzing, which is hydrogen formation as shown in Equation (27-1), stops. This should take about 5 minutes. Remove the beaker from the hot plate and allow it to cool.
5. Add 20 mL of 3 M sulfuric acid (H_2SO_4) **(CAUTION: Corrosive!)** to the beaker and stir the mixture well to dissolve as much of the gel-like material as possible.
6. Suction filter the mixture using a Büchner funnel as your instructor will demonstrate. Be sure that the filter paper lies flat on the bottom of the Büchner funnel. Wet the filter paper slightly with deionized/distilled water to help flatten the filter paper. Take the filtrate (the liquid that went through the filter paper into the flask) and pour it into a 150 mL beaker.
7. Add 10 mL of the filtrate to 10 mL of the dirty water provided in a 100 mL beaker. Add solid sodium hydrogen carbonate ($NaHCO_3$) slowly, with stirring; until a piece of red litmus paper tests blue. [Use your stirring rod to add a drop of the solution to the red litmus paper for this test.]. Stir well and pour some of the slurry (semi-liquid mixture with particles floating in it) into a centrifuge tube and label it tube "1."

8. The aluminum hydroxide gel will settle out over time with its trapped "dirt," but we can speed up the process by using a centrifuge. Using untreated dirty water, fill a second centrifuge tube (tube "2") to the same level as tube "1." Put the tubes in the centrifuge on opposite sides and centrifuge them for 2 minutes. **CAUTION: Do not try to stop the centrifuge with your hands. Just wait until it stops. The centrifuge must be balanced with tubes with equal amounts of contents on opposite sides.**
9. Decant the liquid from the centrifuge tubes into spectrophotometer tubes labeled "1" and "2." Fill a third spectrophotometer test tube (tube "3") with untreated dirty water to the same height as the other two tubes. Compare the absorbance at 500 nm of each solution using a spectrophotometer. Record the results on the lab report form.

   Place the waste from this part of the experiment into the waste container provided.

II. Use of Activated Charcoal

1. Take two Erlenmeyer flasks of the same size and add 50 mL of deionized/distilled water to each flask. Add five drops of methylene blue dye solution to each flask.
2. Add about 0.5 g of activated charcoal to one of the flasks. Swirl the flask to mix the water and charcoal well. This will absorb impurities and contaminants. The water should look better.
3. Filter the charcoal from the water using the suction filtration method used earlier and compare the appearance of the treated water with the untreated water that still has methylene blue in it. Record the results on the lab report form.

The liquid waste from this part of the experiment should be placed in the container designated by your instructor.

EXERCISE 27 PURIFYING CLOUDY WATER – A TREATMENT PROCESS

Pre-Lab Questions

Last Name ____________________ **First Name** ____________________
Instructor ____________________ **Date** ____________________

1. Why should solids be removed from a drinking water source?

2. What is the type of funnel used in this experiment?

3. In Part I, when the solution containing aluminum and sodium hydroxide is heated under the hood, why should you wait until the "fizzing" stops?

4. In Part I, item #8, what is the purpose of the centrifuge?

5. Why would it not be safe for you to drink the water from today's experiment, even if all the steps were followed exactly?

EXERCISE 27 PURIFYING CLOUDY WATER – A TREATMENT PROCESS

Data

Last Name ____________________ **First Name** ____________________
Instructor ____________________ **Date** ____________________

I. Preparation and Use of Aluminum Hydroxide

Compare the volume of the aluminum hydroxide produced (Part I, Step 4) to the original volume of aluminum metal.

____

____

Absorbance (at 500 nm) of test tube with clarified water (tube "1")

____

Absorbance (at 500 nm) of test tube with centrifuged water (tube "2")

____

Absorbance (at 500 nm) of tube with untreated dirty water (tube "3")

____

II. Use of Activated Charcoal

Appearance of test tube treated (with charcoal) dyed water

____

Appearance of test tube untreated (without charcoal) dyed water

____

III. Questions

1. What were the bubbles made of which formed when the aluminum foil reacted with the sodium hydroxide (Part I, Steps 2 and 3)?

2. What were the bubbles made of which formed when the sodium hydrogen carbonate was added to the sulfuric acid solution (Part I, Step 7)?

3. Suppose the activated charcoal absorbed a very toxic substance (but not heavy metals). What would you recommend be done with the used charcoal?

4. Is the clarified water you prepared in Part I now safe to drink? Why or why not?

EXERCISE 27 PURIFYING CLOUDY WATER – A TREATMENT PROCESS

Post-Lab Questions

Last Name ____________________ **First Name** ____________________
Instructor ____________________ **Date** ____________________

1. What are typical suspended solids in a public water source?

2. What is the difference between suspended solids and material dissolved in water?

3. What is the purpose of using activated carbon in a public water treatment process?

4. What is the chemical name for baking soda?

5. What is the function of flocculants in a water treatment process?

BIBLIOGRAPHY

Beard, James M., *Environmental Chemistry in Society*, Second Edition, Taylor & Francis, Boca Raton, FL, 2013, p. 268, 327.

EXERCISE 28

Is This Air Okay to Breathe? A Laboratory Analysis

Most of the soluble salt from urban air is sodium chloride.

Figure 28.1 Fresh air is necessary for health as well as for esthetics.

INTRODUCTION

Air quality is a serious problem in many parts of the world. Fresh air like that shown in Figure 28.1 is not only esthetically pleasing, it is a necessity for good health. In this lab, you are studying the air in an area of automobile

DOI: 10.1201/9781003006565-31

traffic. A beaker of very pure water has been left outside in such an environment for 7 days.

Pollutants from the air can be found in the beaker. Those that dissolved in the water are called soluble solids. Those that did not dissolve in the water are called insoluble solids. The lab involves separating the insoluble solids from the liquid by filtration. Most of the soluble salt from urban air is sodium chloride. The presence of chlorides can be proven by the silver nitrate test. Here is the net equation. Nitrate ion, being a spectator ion in the reaction, does not appear in the equation.

$$Ag^+(aq) + Cl^-(aq) \rightarrow AgCl(s) \qquad (28.1)$$

Certain gases in the air such as CO_2, NO_2, and SO_2 can react with water as shown below.

$$CO_2(g) + H_2O(l) \rightarrow H_2CO_3(l) \qquad (28.2)$$

$$NO_2(g) + H_2O(l) \rightarrow HNO_3(l) \quad \text{(not balanced)} \qquad (28.3)$$

$$SO_2(g) + H_2O(l) \rightarrow H_2SO_3(l) \qquad (28.4)$$

All three of these oxides are formed from non-metals and as such are acid anhydrides. That means in the presence of moisture such as humidity or humans' respiratory systems, they form acids. Just like adding water to instant coffee forms coffee, adding water to acid anhydrides forms acids. It is a principle of chemistry that acids catalyze (speed up) the breakdown of protein. In this case, lung disease and other problems can result.

INSTRUCTIONS

Safety. Put on your PPE.

Supplies for 24 students or teams of students. 24 long-stemmed conical funnels, 24 ring stands, 24 rulers, fine pore filter paper, 12–24 hot plates, 48 250 mL beakers, 24 watch glasses, 6 balances to 0.01 g or better, 24 100 mL graduated cylinders, 6 M nitric acid in dropper bottle CAUTION: Corrosive and contact yellows skin, 0.1 M silver nitrate in dropper bottle (corrosive and contact darkens skin), 24 glass rods, broad range pH paper, 24 test tubes, oven, 24 1.0-L beakers which were placed outside in an area of automobile traffic for a week, 24 metric rulers. These beakers originally were about one-half full with deionized/distilled water.

Additional water was added if needed to replace that lost by evaporation. Ethanol can be added as antifreeze if needed. In this case, the beakers need to be in a secured area away from children and pets.

Determination of Insoluble Solids. Weigh a piece of fine pore filter paper and record this weight on line 2 of Table 28.1. Fold the paper in half, and then fold it again, almost in half. Open it so that three layers of paper are on one side and one layer on the other. Place it carefully in the funnel supported by a ring stand as shown in Figure 28.2.

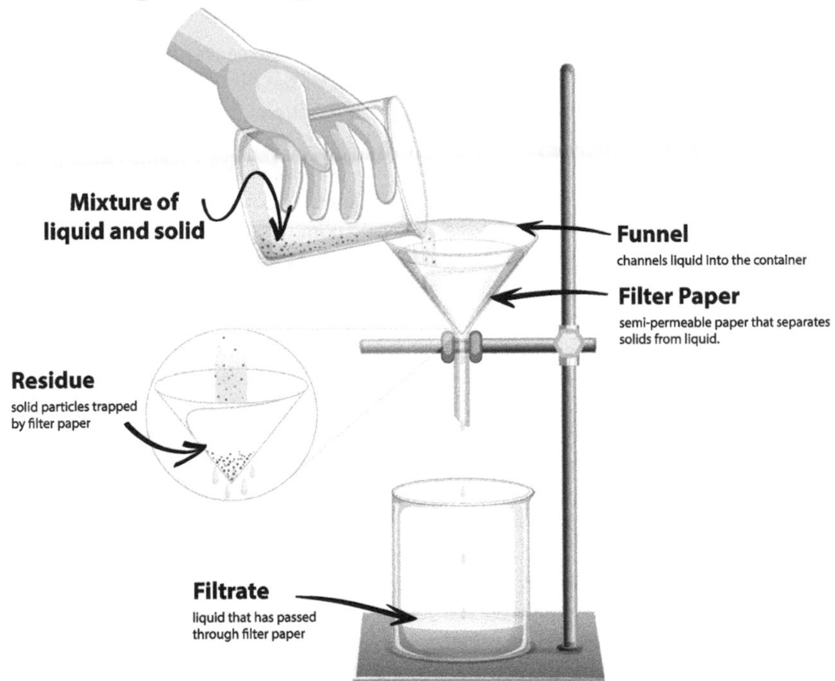

Figure 28.2 The filtration process explained.

Taking time to ensure that the filter paper adheres closely to the funnel, and adding a few drops of deionized/distilled water if necessary to accomplish this, can save time in the filtration process. Filter your water sample through the weighed filter paper, keeping the filtrate (liquid that comes through the funnel) in a collection beaker for further tests. Place the filter paper and contents on a clean watch glass and dry in an oven at about 105°C. Allow the filter paper to cool to room temperature.

[Without cooling objects to be weighed, the balance can be damaged and the weight can be in error due to convection currents – heat rises!] Weigh the dried filter paper and record the weight on line 1 of Table 28.1. Calculate the weight of insoluble solids picked up and record on line 3 of Table 28.1.

Measure the diameter of the beaker used in cm, and record on line 4 of Table 28.1. Calculate the area of the beaker in cm^2 and record on line 6. Convert the area to square meters and record on line 7. Record on line 8 grams of solid per square meter. Express the results obtained in grams of solid per square meter per day. Calculate grams of solid per square meter per day by dividing line 8 by 7 (days). Record results on line 9.

Area of the water's surface = Π (radius)^2 where radius = ½ (diameter)

Area of the water's surface in m^2 = Area of the water's surface in cm^2 × (1 m^2/10,000 cm^2)

Determination of Soluble Solids. Weigh a 250 mL beaker and record the weight on line 2 of Table 28.2. Place 100 mL of the filtrate collected above, in the 250 mL beaker. Evaporate slowly on a hot plate, to near-dryness. Be sure to remove the beaker from the hot plate when it is almost dry. If you wait until the beaker is dry, it can get too hot and may break; it may also cause your dissolved solids to splatter off and fly around the room, causing error and safety issues. Cool the beaker to room temperature; then weigh it and record the weight on line 1 of Table 28.2 of your data sheet. Subtract line 2 from line 1 to calculate the weight of dissolved solids for line 3. The soluble solids are chiefly sodium chloride. Make a rough approximation of the amount of chloride picked up by dissolving the solid obtained by evaporation in the least possible amount of 6 M nitric acid (HNO_3), stirring carefully with a glass rod. **CAUTION: Nitric acid is corrosive**. Transfer the solution to a test tube and add several drops of 0.1 M silver nitrate ($AgNO_3$). **CAUTION: Silver nitrate is corrosive – avoid contact**. A white precipitate of silver chloride (AgCl) indicates the presence of chloride ions. Record your observations and conclusions on lines 11 and 12 of the data sheet under Table 28.1.

Determination of Acidity (pH). Use broad-range pH paper for the following and remember that a pH of 7 is neutral, below 7 is acidic, and above 7 is basic. Check the pH of deionized water in the lab to answer Question 1 in Table 28.3. Check the pH of the water you received today in the beaker. Subtract. Record results in Table 28.3.

Figure 28.3 Rural areas may offer cleaner air than cities.

EXERCISE 28 IS THIS AIR OKAY TO BREATHE? A LABORATORY ANALYSIS

Pre-Lab Questions

Last Name ________________ **First Name** ________________
Instructor ________________ **Date** ________________

1. Name the two types of solids for analysis in this lab.

2. How will you test for acid anhydrides and acid rain in your sample?

3. Name two corrosive compounds used in this lab.

4. Write the equation for the chloride test in this lab.

5. What separation method is used in this lab?

EXERCISE 28 IS THIS AIR OKAY TO BREATHE? A LABORATORY ANALYSIS

Data

Last Name ____________________ **First Name** ____________________
Instructor ____________________ **Date** ____________________

Table 28.1 **Determination of particulate solids in air**

Number	Data to Collect	Your Results
1	Wt. of filter paper + insoluble solids (g)	
2	Wt. of filter paper (g)	
3	Wt. of insoluble solids (g)	
4	Diameter of beaker (cm)	
5	Radius of beaker (cm)	
6	Area of beaker (cm^2)	
7	Area of beaker (m^2) [To get this, divide your answer to 6 above by 10,000.]	
8	Grams of solid/(m^2). [To get this, divide your answer to 3 above by your answer to 7.]	
9	Grams of solid/(m^2) per day. [To get this, divide your answer to 8 by the number of days the beaker was left outside.]	

10. Interpret your results from part 9 above, keeping in mind that in general, soot fall should not exceed 0.65 g/m^2 in residential areas. What is your assessment of the quality of the air you just tested?

Table 28.2 **Determination of dissolved solids in air**

Number	Data to Collect	Your Results
1	Wt. of 250 mL beaker + dissolved solids (g)	
2	Wt. of empty 250 mL beaker (g)	
3	Wt. of dissolved solids (g)	

11. Describe your observations when the nitric acid was added to the residue.

12. Does this indicate the presence of chloride ions?

Table 28.3 Determination of pH of water exposed to air

Number	Data to Collect	Your Results
1	pH of deionized water first placed in the beaker which was exposed to the air	
2	pH of water which was exposed to air for 5 days	
3	Change in pH (2-1)	

4. The pH
   a. stayed the same
   b. decreased
   c. increased.
   Any increase in acidity is due to CO_2, SO_2, and NO_2, formerly in the air, which dissolved in the water.

5, 6, and 7. Write the reaction for each of the above with water in the form of a balanced equation.
   a. $CO_2 + H_2O \rightarrow$ ______________ Predicted pH of solution is acidic, basic, or neutral?
   b. $SO_2 + H_2O \rightarrow$ ______________ Predicted pH of solution is acidic, basic, or neutral?
   c. $NO_2 + H_2O \rightarrow$ ______________ Predicted pH of solution is acidic, basic, or neutral?

EXERCISE 28 IS THIS AIR OKAY TO BREATHE? A LABORATORY ANALYSIS

Post-Lab Questions

Last Name ____________________ **First Name** ____________________
Instructor ____________________ **Date** ____________________

1. Were you surprised by the results of this air quality lab? If so, how?

2. Do these results influence where you want to live, build a house, or jog in the future? Please comment.

3. What do you think about senior citizen homes and hospitals that are built on or near busy freeways, considering that many of their residents are already in frail health?

4. What choices besides moving can we make to protect ourselves from breathing contaminated air?
   a. Staying indoors during times of heavy pollution such as the use of fireplaces
   b. Asking our elected representatives to vote in favor of cleaner air measures
   c. Installing air purifiers in our homes and offices
   d. All of the above
5. Chemically speaking, why are carbon dioxide, sulfur dioxide, and nitrogen oxides harmful to breathe if in excessive quantities? What type of compound are they?

EXERCISE 29

What Are Plastics, Anyway? A Study of Polymers

Have we been too successful with plastics, considering their unwanted presence in the environment these days, even in oceans?

Figure 29.1 An elephant parade is an excellent model of a polymer's structure.

INTRODUCTION

Think of a *long* line of elephants in a circus parade, connected trunk to tail (Figure 29.1); then replace the elephants with ethylene ($CH_2{=}CH_2$) molecules and you have the polymer, polyethylene. Replace the elephants with

DOI: 10.1201/9781003006565-32

tetrafluoroethylene (CF_2=CF_2), and you have the polymer Teflon™. Molecules that join to form polymers, such as ethylene, are called **monomers**; often these monomers have carbon-carbon double bonds that make this reaction possible. Polyethylene and Teflon™ are examples of addition polymers. Other types of polymers are formed by molecules which release water as they join, and are called condensation polymers. Since there are many molecules which can assemble themselves in this manner, the possibilities for forming these long molecules (aka polymers or plastics) are endless. Have we been too successful with plastics, considering their unwanted presence in the environment these days, even in oceans? Or is it that we just need to be more responsible in using (and disposing of) polymers? And remember that we ourselves are made in part of polymers such as glycogen, the emergency energy starch found in our liver; DNA which holds our heredity; and proteins which give us muscle, hair, and nails, to name a few.

In this four-part lab, you are to study starch pellets, sodium polyacrylate, Amazing Enviro-Bond™, and a cross-linked polymer.

Packing material is used to protect items that we purchase during storage, shipping, and handling. An example is Styrofoam™ (polystyrene) which is used to protect everything from drinking glasses to computers. It prevents millions of dollars' worth of damage each year. Styrofoam™ is also used to make disposable food and beverage containers. The disposal of the Styrofoam™ presents a problem, in that this material is **non-biodegradable** (bacteria cannot break it down into its components) and it occupies a large volume of landfill space. It is possible to recycle the Styrofoam™, but few recycling centers will accept it because of the problems associated with transporting it to polystyrene recycling plants.

Some manufacturers and shippers are using an alternative material, starch, to make packing pellets. Starch can be processed to include many of the same foam-like properties of Styrofoam™, making it a good packing material. It has the advantage of being **biodegradable** (bacteria *can* break it down into its components) since the primary ingredient is plant-based such as corn. Crops like these, produced agriculturally, are thus **renewable** resources, which means that they can be replaced. Styrofoam™ is produced from petroleum products. Petroleum is a **non-renewable** resource; once we have used it all, it cannot be replaced. The major disadvantage of of these agriculturally based products is increased production cost compared to that of Styrofoam™.

In Part 1 of today's lab exercise, we are to examine the properties of starch pellets. A common chemical test for starch is the addition of (yellow-brown) iodine, which forms an inky-blue complex to indicate the presence of starch, such as one would find in potatoes or corn.

In Part II you assume the role of a chemist in a Research and Development (R&D) industrial lab and your supervisor has assigned your group to study a "new" material, sodium polyacrylate.

Part III addresses the concern of spills of crude oil (unprocessed hydrocarbon mixtures from underground geologic formations) which occur all too often. These spills are especially problematic on water where they form an oily top layer as they are less dense than water. An example of such a disaster is the Exxon Valdez spill that occurred near the environmentally sensitive Alaskan shore. Millions of dollars spent on this clean-up did not prevent considerable environmental damage. *Better* clean-up methods were suggested as a remedy for future such problems, so oil companies continually seek better ways to contain and clean up oil spills.

Enviro-Bond™ is a polymer that was developed to help with this problem. It has been approved for this use by the Environmental Protection Agency (EPA). In Part III of our lab exercise, we are to examine the effect of this polymer on a mixture of oil and water.

Part IV of the lab exercise starts with a polymer already dissolved in water. On a molecular scale, as shown in Figure 29.2, a polymer could look like the drawing on the upper left. If we add a second material, we can cause the chains in the original polymer to form new bonds or links between them. This process is called "cross-linking" a polymer and is also shown in Figure 29.2, in the lower left structure. You will notice a big difference between the original polymer and the final cross-linked polymer. This type of linkage is present in the proteins in our bodies, as the giant molecular structures are stabilized in part by cross-links involving sulfhydryl (-SH) groups. The structures on the right in Figure 29.2 show images of some additional types of polymers. Here we are focusing on cross-linked polymers.

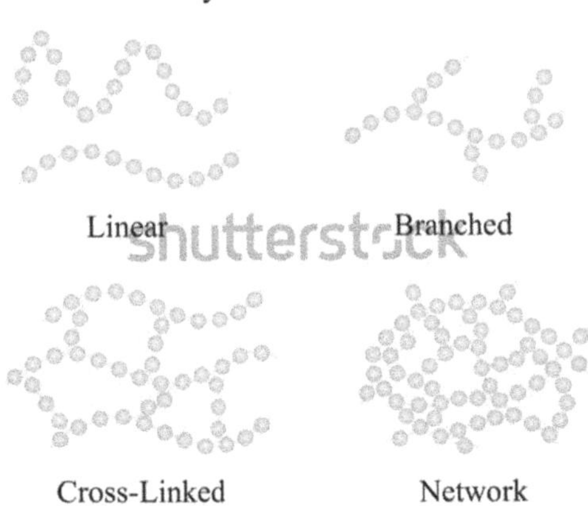

Figure 29.2 Types of polymers.

INSTRUCTIONS

Safety. Put on your PPE.

Supplies for 24 students or teams. 96 250 mL beakers, 50 pieces of starch pellets, hot water or 12 hot plates, 50 mL tincture of iodine, 96 250 mL beakers, marker or grease pens to label beakers, 96 glass stirring rods, 24 100 mL graduated cylinders, food coloring, 6 balances to 0.1 g or better, weighing boats, sodium chloride (non iodized table salt), 12 heat-resistant mitts or beaker tongs, gloves, food coloring, weighing boats (blue or other color recommended for contrast), 300 g sodium polyacrylate, deionized/distilled water, tap water, 24 ziplock plastic bags, 300 mL Marvel Mystery Oil™, 24 g EnviroBond™, 24 2 oz. plastic cups with lids, 4-24 10 mL graduated cylinders, 4-24 25 mL graduated cylinders, 600 mL 4% aqueous polyvinyl alcohol, 100 mL 4% aqueous sodium borate (Borax) solution.

PART I STARCH PELLETS

Step 1. Add about 200 mL of hot water from the hot water tap or, if not available, heat water in a beaker with a hot plate. **CAUTION: Use heat-resistant mitt or beaker tongs to avoid burning your hands**. Drop two starch pellets into the hot water and stir with a stirring rod. Describe what happened (Part I, Question 1).

Step 2. Add 4 drops of tincture of iodine to the mixture of starch pellets and water and stir. The solution may be disposed of down the drain in many areas. Describe and explain what happened (Part I, Question 2); also answer Part I, Questions 3–4.

PART II SODIUM POLYACRYLATE

Step 1. Describe the physical state (solid, liquid, or gas) and color of the sodium polyacrylate. Record under Part II, Question 1 of your data page.

Step 2. Complete the following process.

1. Number four 250 mL beakers from 1 to 4 using a grease pencil or marker.
2. Using a graduated cylinder, measure and pour 200 mL of deionized water into each of the four beakers.
3. Add two drops of food coloring to the water in each beaker. Choose *one* color for all four beakers.
4. Weigh out the following three samples of sodium chloride (table salt) into weighing boats: (1) 2.0 g, (2) 1.0 g, and (3) 0.5 g. CAUTION: Never consume anything in lab! Even a familiar household item like table salt, when in a lab, could have contaminants.
5. Add the 2.0-g sample of salt to beaker 1, stirring until the salt is completely dissolved.

6. Add the 1.0-g sample of salt to beaker 2, stirring until the salt is completely dissolved.
7. Add the 0.5-g sample of salt to beaker 3, stirring until the salt is completely dissolved. Do not add salt to beaker 4.
8. Weigh 2.0-g samples of sodium polyacrylate into each of four weighing boats.
9. Stir the water in beaker 1 *slowly* while your lab partner quickly pours 2.0 grams of sodium polyacrylate into the water. Continue stirring for approximately 1 minute. Record your observations (Part II, Question 2a).
10. Stir the water in beaker 2 slowly while your lab partner quickly pours in 2.0 g of sodium polyacrylate. Continue stirring for 1 minute. Record your observations (Part II, Question 2b).
11. Stir the water in beaker 3 slowly while your lab partner quickly pours in 2.0 g of sodium polyacrylate. Continue stirring for 1 minute. Record observations (Part II, Question 2c).
12. Stir the water in beaker 4 slowly while your lab partner quickly pours in 2.0 g of sodium polyacrylate. Continue stirring for 1 minute. Record observations (Part II, Question 2d).
13. Answer Questions 3–5 in Part II.

PART III AMAZING ENVIRO-BOND™

Step 1. Measure 200 mL of tap water with a graduated cylinder, and add it to the ziplock bag.

Step 2. Add 10 mL (2 teaspoons) of Marvel Mystery Oil to the water in the bag. Close the bag tightly and shake carefully, ensuring that none of the mixture leaves the bag. Describe the contents (Part III, Question 1); also answer Question 2.

Step 3. Measure 1.0 g of Enviro-Bond™ into a weighing boat with the balance. Describe this polymer as to physical state, appearance, and color (Part III, Question 3).

Step 4. Add the Enviro-Bond™ you weighed out in Step 3 to the bag, close carefully, seal, and shake the bag ensuring no spills occur. Describe the results and tell how the oil and water might be separated now (Part III, Questions 4–5). Discard the sealed bag into the designated waste bin – not sink or trash.

PART IV CROSS-LINKED POLYMER

Step 1. Measure *only* 15–20 mL of the 4% aqueous polyvinyl alcohol (PVA) solution in a 25 mL graduated cylinder, and transfer it to the 2 oz. cup. Write down its appearance and whether it seems viscous (thick). You may add a drop or two of food coloring, mixing with the stirring rod, to the cup to obtain a colorful polymer!

Step 2. Measure 3 mL of the sodium borate (Borax) solution with the 10 mL graduated cylinder, and add to the PVA solution, stirring quickly and thoroughly. Describe what happens. *Wearing gloves,* remove the gel from the cup and knead it – like modeling clay. Describe how it feels and acts. Hint: if the mixture does not gel, add a few more milliliters of sodium borate solution. Pull it slowly, to see if it will form a thin film. Pull it fast, and see what happens. Make a ball with it and see if it will bounce.

EXERCISE 29 WHAT ARE PLASTICS, ANYWAY? A STUDY OF POLYMERS

Pre-Lab Questions

Last Name ____________________ **First Name** ________________
Instructor ____________________ **Date** ________________

1. Name two common uses of Styrofoam in our everyday life.

2. What is the environmental problem with these Styrofoam uses?

3. What is the name of a relatively new packing material studied in this experiment which is a possibly superior alternative to Styrofoam?

4. What is the environmental advantage in the use of this new material?

5. What is this new material's primary ingredient?

6. What is your investigation of sodium polyacrylate supposed to show?

7. How should you dispose of the beaker contents after the sodium polyacrylate part of the experiment, when you have finished with them?

8. What environmental problem was the new polymer EnviroBond™ developed to address?

9. Where in the lab do you dispose of the ziplock bag and contents used in the above experiment?

10. Define the term "cross-linking" of a polymer.

EXERCISE 28 WHAT ARE PLASTICS, ANYWAY? A STUDY OF POLYMERS

Data

Last Name ____________________ **First Name** ____________________
Instructor ____________________ **Date** ____________________

Part I. Starch Pellets

1. Describe what happened with the hot water.

2. Describe what you *saw* with the iodine test.

3. Explain what the iodine test results mean.

4. Explain how we know that starch pellets are biodegradable.

Part II. Sodium polyacrylate

1. Describe this mystery material.

2. Record your observations for
   a. Beaker 1 ____________________
   ____
   b. Beaker 2 ____________________
   ____
   c. Beaker 3 ____________________
   ____
   d. Beaker 4 ____________________
   ____

3a. Do you think this mystery material could be, based on one of its properties, commercially important? Explain. Hint: Think about what happened when the substance was placed in water.

3b. What property of the mystery material seems to make it of commercial importance?

3c. List three or more possible uses for this mystery material.

4. State the effect of salt (sodium chloride) on the property you listed in Question 3b.

5. If your instructor authorizes you to pour the contents of all four beakers down the drain, what must be done to their contents? Hint: this is especially important for beaker 4.

Part III. Amazing Enviro-Bond™

1. Describe the contents. Did they mix? Explain where the oil is.

2. How might you clean up an oil spill without adding chemicals?

3. Describe the Enviro-Bond™ as to color, physical state (solid, liquid, or gas), and appearance.

4. Describe what you saw when the Enviro-Bond™ was added to the mixture in the bag. What clue does this possibly provide as to a way to separate oil and water?

5. Does Enviro-Bond™ look promising for cleaning up oil spills in the oceans? Why or why not?

6. Compare the properties of Enviro-Bond™ to those of sodium polyacrylate. How are they similar and how are they different?

Part IV. A cross-linked polymer

1. General appearance and thickness of polymer solution.

2. Results when Borax was added.

3. Polymer properties (what it feels like and how it behaves).

4. Did the polymer make a thin film? Describe what you saw.

5. What happened when the polymer was quickly pulled?

6. When shaped into a ball, did it bounce? Describe what you saw.

EXERCISE 29 WHAT ARE PLASTICS, ANYWAY? A STUDY OF POLYMERS

Post-Lab Questions

Last Name ____________________ **First Name** ____________________
Instructor ____________________ **Date** ____________________

1. Given that polyethylene (Figure 29.3) is a polymer of ethylene, polypropylene must be a polymer of ______________________________.
2. Given that the chemical name of Teflon™ (Figure 29.4) is polytetrafluoroethylene, why do you think it is more commonly called Teflon™?
   ____
   ____
3. List five common objects in your home that are made of plastics.
   ____
4. List two advantages of plastics.
   ____
   ____
5. List two disadvantages of plastics.
   ____
   ____

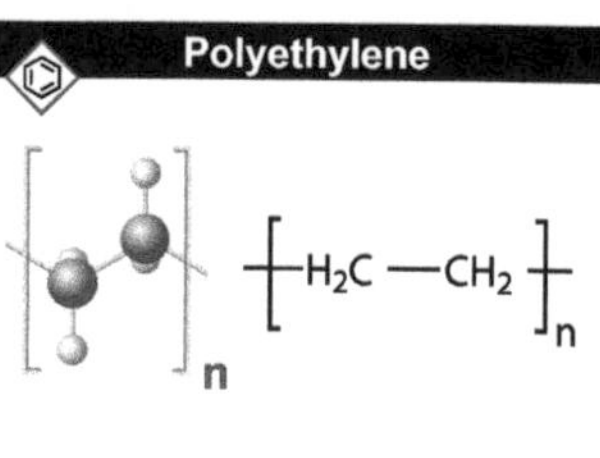

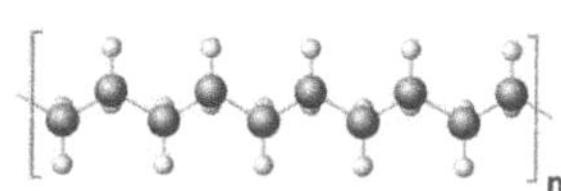

Figure 29.3 Structure of polyethylene, a common plastic.

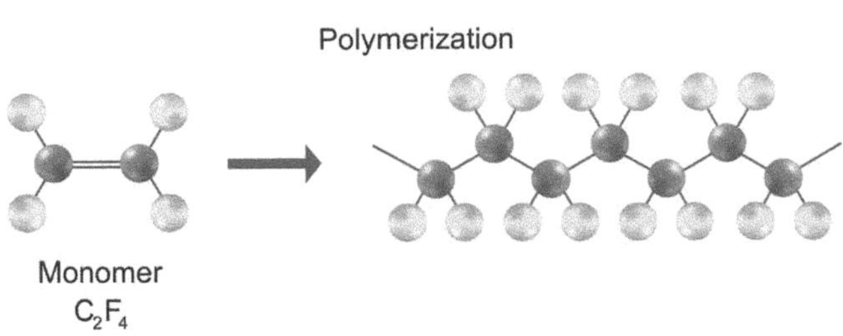

Figure 29.4 The polymerization reaction leading to Teflon™.

EXERCISE 30

Should I Bother with Recycling? Separation of Plastics by Density

Which is heavier, a ton of lead, or a ton of corks?

Figure 30.1 Lead, with an atomic mass of 207 g/mole, is known as a heavy element.

INTRODUCTION

Which is heavier, a ton of lead, which is shown in Figure 30.1, or a ton of corks, which are shown in Figure 30.2? The answer to this old question is "neither." Many people confuse the terms "heavy" and "dense." "Heavy"

DOI: 10.1201/9781003006565-33

refers to mass, and "dense" refers to mass per unit volume. The lead has the greater **density** since the 40 g of lead will have only a volume of 3.54 mL, while 40 g of corks will have a volume of 190.4 mL.

The mass of lead contained in 1.0 mL is its density:

Density of lead = Mass/Volume = 40 g/3.54 mL = 11.3 g/mL

The mass of cork contained in 1.0 mL is its density:

Density of cork = Mass/Volume = 40 g/190.4 mL = 0.21 g/mL

Since the mass of 1.0 mL of water at 4°C is exactly 1.0 g and changes only slightly with temperature, we will assume that the density of water at room temperature is 1.0 g/mL. Also, 1.0 mL is exactly 1.0 cubic centimeter (cc) so these units can be used interchangeably. An object with a density greater than 1.0 g/mL will *sink* in water and an object with a density less than 1.0 g/mL will *float* in water.

Before plastics can be recycled they must be separated into one of the seven numbered recycling groups as shown in Figure 30.3. Although there are many types of plastics, the most common element in them is carbon. Carbon has the amazing ability to bond with both metals and non-metals. It can also bond with itself, forming long *chains* of carbon atoms, a property called **catenation**. Plastics are polymers, as explained in Exercise 29, having long chains of carbon atoms, with various functional groups which modify their properties. Since they contain much carbon, they are frequently formed from non-renewable hydrocarbons (petroleum), and they are frequently flammable.

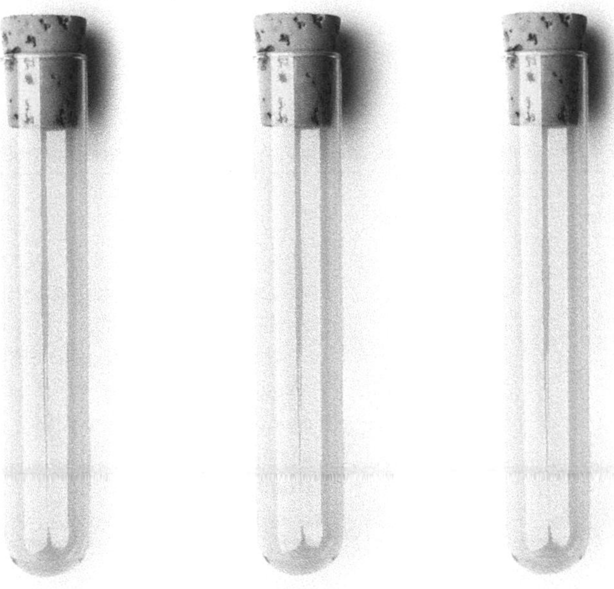

Figure 30.2 Corks are very light and float in water; thus, they are less dense than water.

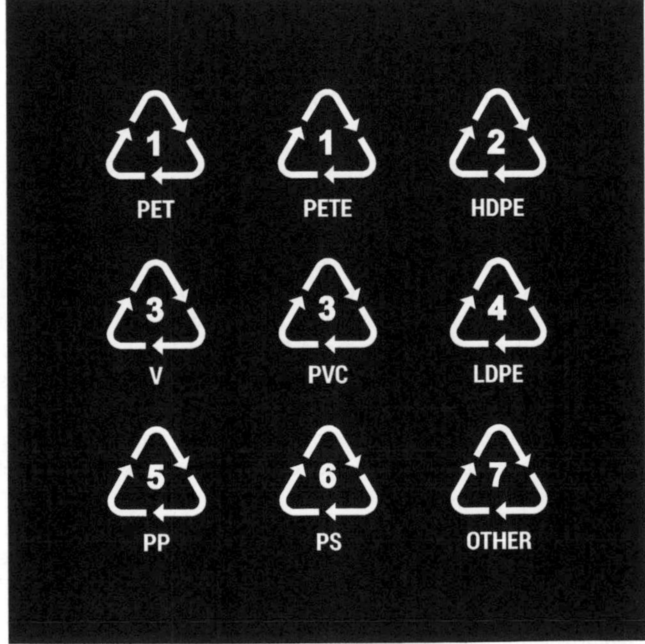

Figure 30.3 Classifications of plastics (polymers) for recycling purposes.

INSTRUCTIONS

This separation can be done in the lab by comparing the relative densities of plastics, as shown in Table 30.1.

Table 30.1 Classification of plastics by their number and density

Plastic Number	Density (g/cc)	Name
1	1.35–1.38	Polyethylene Terephthalate (PETE)
2	0.94–0.96	High-Density Polyethylene (HDPE)
3	1.32–1.42	Polyvinyl Chloride (PVC)
4	0.91–0.93	Low-Density Polyethylene (LDPE)
5	0.90–0.92	Polypropylene (PP)
6	1.03–1.06	Polystyrene or Styrofoam (PS)
	1.20	Polycarbonate
	1.202 g/mL	Saturated Sodium Chloride
	0.930 g/mL	Isopropyl Alcohol Solution
	0.917 g/mL	Corn Oil
7		Other

Safety. Put on your PPE.

Supplies for 24 students or teams. Deionized/distilled water, 96 250 mL beakers or plastic cups, plastics samples of various groups, 24 pairs of forceps, paper, 24 100 mL graduated cylinders, 1800 mL 100% isopropyl alcohol, red and green food coloring, 24 glass stirring rods, 2.4 L corn oil, 2.4 L saturated sodium chloride (NaCl) solution.

1. Fill a plastic cup or 250 mL beaker about half full with deionized/distilled water.
2. Record the sample number of your unknown on our data page. The unknown will include various assorted plastics. Place all of these pieces in the water prepared in step 1, and stir well with your glass stirring rod. Tap the container gently on your desk to dislodge any air bubbles attached to the plastic. These can cause error as air has a density different from plastic's. The plastics with densities less than that of water, which is 1 g/mL, will float, while those with a density greater than water's will sink.
3. Using forceps remove the *floating* pieces of plastic from the water and place them on a paper labeled "less dense than water."
4. Pour off most of the water; then remove the rest of the plastic pieces from the bottom of the cup and place them on a piece of paper labeled "denser than water."
5. Use a graduated cylinder to measure 45 mL of 100% isopropyl alcohol ($CH_3CHOHCH_3$) and add this to a 250 mL beaker; then measure 55 mL of deionized/distilled water and add it to the beaker, along with a drop of red food

coloring. Mix carefully with a glass stirring rod.

CAUTION: Isopropyl alcohol is flammable. No flames should be in the room. The solution also irritates the eyes and skin. GOGGLES MUST BE WORN! Rinse off if it gets on your skin.

6. Add only the plastic pieces which floated in water (from Step 3) to the freshly prepared alcohol solution. Stir well. Again, gently tap the container on the desk to dislodge any air bubbles from the plastic.
7. Remove any floating pieces of plastic using forceps and place on paper labeled "less dense than alcohol solution."
8. Remove the sinking pieces of plastic and describe their density. Identify the plastic.
9. Place the *floating* pieces from Step 7 in a cup or beaker containing about 100 mL of corn oil and stir. Again, remove any air bubbles adhering to the plastic. This step can take awhile because of the viscosity of the corn oil. After no more motion on the part of the plastic pieces is apparent, identify the floating pieces and the sinking pieces.
10. Describe the density of the plastics in Step 9.
11. Pour approximately 100 mL of saturated salt solution in a cup or beaker. Add one drop of green food color to identify it.
12. Take the pieces of plastic which sank to the bottom of the deionized water in Step 4 and place them in the saturated salt solution. Stir carefully. Tap gently on desk to dislodge any air bubbles.
13. Using forceps, separate the pieces which float from the pieces which sink. Describe the density of each of the plastics. Identify the plastics using Chart 1 as a reference.

EXERCISE 30 SHOULD I BOTHER WITH RECYCLING? SEPARATION OF PLASTICS BY DENSITY

Pre-Lab Questions

Last Name ____________________ **First Name** ____________________
Instructor ____________________ **Date** ____________________

1. Which is heavier, a ton of steel or a ton of feathers? Explain.

2. Explain the difference between "heavy" and "dense."

3. What is the density of water at room temperature?

4. Objects with a density *greater* than water will ______________ in water.

5. Objects with a density *less* than water will ______________ in water.

6. Which of the answers to Questions 4 and 5 would High-Density Polyethylene (HDP) do in water?

7. What alcohol will we be using to study the separation of plastics? Write the chemical formula.

8. What precautions should be observed while using this alcohol?

9. Aside from water and alcohol, what other liquids will we use to observe the separation of plastics?

10. How does Table 30.1 help identify your plastic unknowns?

EXERCISE 30 SHOULD I BOTHER WITH RECYCLING? SEPARATION OF PLASTICS BY DENSITY

Data

Last Name __________________ **First Name** __________________
Instructor __________________ **Date** __________________

1. Sample # ______________________________

2. How many pieces of plastics from your sample floated in water?
____

3. How many sank in water? ______________________________

4. What does this tell you about their densities?
   a. Floating pieces ______________________________
   b. Sinking pieces ______________________________

5. How many pieces of plastic floated in the isopropyl alcohol solution?
____
   a. What can you say about their densities? ____________________

6. How many pieces of plastic sank? ____________________
   a. What can you say about their densities? ____________________
   b. To which recycling group do these pieces belong? (Give number and names.)

   ____________________ ____________________
   Number Name

7. How many of the pieces of plastic floated in the corn oil? __________
____
   a. What can you say about their density? ____________________
   b. Identify this plastic.

   ____________________ ____________________
   Number Name

8. How many pieces of plastic sank in the corn oil? ____________________
   a. What can you say about their density? ____________________
   b. Identify this plastic.

   ____________________ ____________________
   Number Name

9. How many pieces of plastic floated in saturated salt water? ___________
   a. Density of plastic? ___________
   b. Identify this plastic.

   ___________ ___________
   Number Name

10. How many pieces of plastic sank in saturated salt solution? ___________
    a. Density of plastic? ___________
    b. Identify this plastic.

    ___________ ___________
    Number Name

EXERCISE 30 SHOULD I BOTHER WITH RECYCLING? SEPARATION OF PLASTICS BY DENSITY

Post-Lab Questions

1. What is the most common chemical element in a plastic?

____

2. Why is recycling of plastics important?

3. Density of plastics is determined by what two values?
   a. ____
   b. ____

4. Describe how the relative densities of plastics was determined in today's experiment.

5. Elemental liquid mercury (Hg) has a density of 13.6 g/mL. Based on this, why do you think it was mistakenly thought that waste mercury could safely be dumped into lakes and oceans?

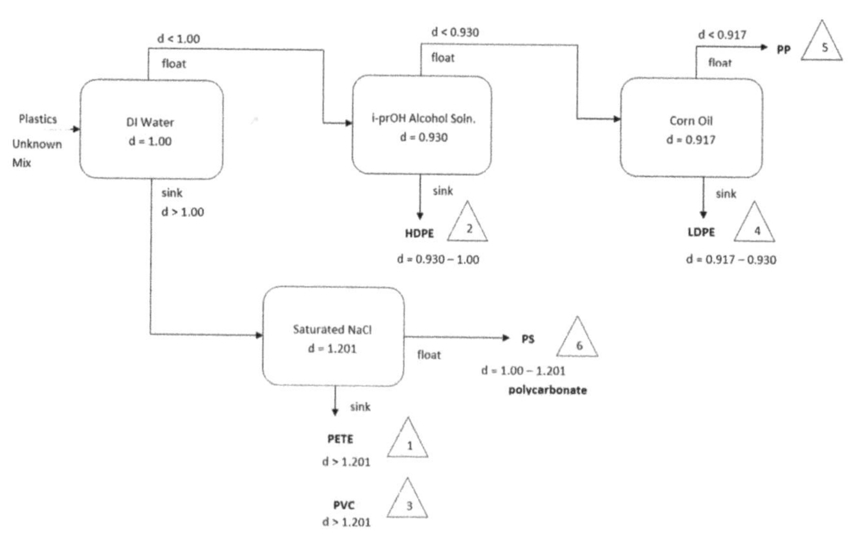

Chart 1 Flow chart for separation of plastics based on densities.

BIBLIOGRAPHY

http://www.teachingplastics.org/hands_on_plastics/activities/plastics_analysis_lab/a1.html

Keenan, Charles and Wood, Jesse, *General College Chemistry*, Second Edition, Harper & Bros, New York.

UNIT C

Closing Statements for the Environmental Chemistry Lab

EXERCISE 31

What Is Illegal about Pollution? An Environmental Law Presentation

In addition to natural laws like the Law of Conservation of Energy, Environmental Chemistry also deals with governmental laws - at local, state, and federal levels.

Figure 31.1 Laws are written and enforced to protect the environment.

INTRODUCTION

Earlier exercises in this text deal with chemical principles and their effect on the environment. As an example, **diffusion** is the process in which a chemical moves from a *more* concentrated to a *less* concentrated area. This means that pollutants at one site, if not properly managed, will spread out into other areas. Just picture a smokestack producing smoke, which drifts over the country side. Or imagine waste hot water from a nuclear power plant being released into a river where the rapid temperature change kills the marine life. To protect citizens from loss of health and damage to property, and to protect environmental quality in general, governments create laws to require proper care of the environment. The enforcement of these laws is then delegated to specified agencies.

Environmental legislation can fill many books like those shown in Figure 31.1, and can be difficult to understand and interpret. As we can well imagine, corporations would prefer to hire professionals who know both

DOI: 10.1201/9781003006565-35

science and the law, to guide them in adhering to environmental legislation, than to be cited for non-compliance with environmental laws. Some infractions of these laws are designated as environmental crime. Penalties for non-compliance can range from hefty fines to prison sentences. The following presentation was contributed by Mr. John C. Mayfield who is shown in Figure 31.2, a chemist employed by the Aluminum Corporation of America (ALCOA) to help ensure their proper compliance with governmental policy. He helped manage environmental concerns for a chemical plant such as that shown in Figure 31.3. We can learn much from his account.

Figure 31.2 Mr. John Mayfield, M.S., ALCOA executive, retired, University of Mary Hardin-Baylor faculty, retired.

Figure 31.3 Chemical manufacturing plants are subject to legislation designed to protect the workers, the population in the surrounding area, and the environment.

INSTRUCTIONS

EXERCISE 31 WHAT IS ILLEGAL ABOUT POLLUTION? AN ENVIRONMENTAL LAW PRESENTATION

Last Name____________________ **First Name** ____________________
Instructor ____________________ **Date** ____________________

Instructions: View the presentation below and answer the following questions. Your instructor may discuss this material in class.

1. For what corporation did the originator of this material work?

2. How much aluminum oxide per day was produced at the facility?

3. List the seven agencies involved in overseeing the industrial plant.

4. Most federal environmental laws are enforced by which of the above seven agencies?

5. What federal law allows agencies to develop regulations which have the force of law?

6. Name the federal law that regulates the reporting of adverse effects of chemicals.

7. Name the law that defines hazardous wastes, treatment, and disposal requirements.

8. Name the law that established the Superfund Process and which was intended to address existing environmental sites and remediation requirements.

9. Name the law that provides for the identification and communication of the presence of chemicals in local communities.

10. Name the oldest federal law.

11. Prior to 1972 permits to discharge water were issued by

12. Permits to discharge substances into natural waters usually last for _____ years.

13. Failure to comply with government policy for the discharge of substances into natural waters carries
    a. civil penalties
    b. criminal penalties
    c. civil and criminal penalties
    d. none of these

14. The first federal law for air quality was passed in ____________.

15. Facilities do not have to submit reports of their air emissions to the federal government.
    a. True
    b. False

16. The impact of a facility's air emissions can be related to
    a. human health
    b. wildlife
    c. plant life
    d. facilities
    e. all of these

17. Permit limits for emissions to the air are generally based on

18. Environmental management systems for large plants must be

19. The "lime" referred to in this lab has the chemical name

20. The facility discussed in this presentation had __________ (give number) water discharge points.

Extra Credit Discussion Question. Would you consider a career in environmental management, such as described in this lab? Why or why not?

Environmental Laws, Regulations and Management

The information presented today is intended to provide limited highlights of federal environmental laws and regulations developed to implement various federal actions.

Adapted from a Presentation Courtesy of Mr. John C. Mayfield, M.S. Chemistry, Retired ALCOA Executive, Retired University of Mary Hardin-Baylor Faculty

To Put Things in Perspective:

- The industrial facility where I was the Environmental Manager for 23 years had the following operations:
  - Aluminum Smelter (1 million pounds per day)
  - Oil and Gas Operations (drilling, pipelines and refining)
  - Chlorine and Sodium Hydroxide (500 tons per day each)
  - Aluminum Fluoride Production

(Operations, Continued)

- Cryolite (Sodium Aluminum Fluoride)
- Lime Kiln (CaO)
- Aluminum Oxide Production (14.5 million pounds per day)
- 12 Natural Gas Fired Steam/Electric Generators

(Operations, Continued)

- Solid Waste Management Facilities (20 million pounds per day)
- Potable (Drinkable) Water Treatment
- Sanitary Water Treatment
- Fifteen Water Discharge Points
- Air Emissions
- Deep Water Harbor
- 3000 Acre Plant Site
- Employee Headcount Peaked at 2600 +

Federal Agencies and Offices Involved:

- US Environmental Protection Agency
- US Corps of Engineers
- US Fish and Wildlife Service
- US Geological Survey
- US Bureau of Marine Fishers
- National Oceanic & Atmospheric Administration
- US Department of Justice

State Agencies and Offices Involved:

- Texas Commission on Environmental Quality
- Texas Department of Health
- Texas Parks and Wildlife Department
- Texas Historical Commission
- Texas Land Commission
- Texas Attorney General

Federal Environmental Laws

- Highlighted are a selected few of the federal laws involved. The title and date of enactment are listed followed by a very brief description of purpose.
- Most federal environmental laws are enforced by the US Environmental Protection Agency (EPA).
- In some cases, laws related to the environment are enforced by other agencies; for example, dredge and fill permits under Section 404 are administered by the US Corps of Engineers.

Where do federal laws and the resulting regulations come from?

- As we know, the US Congress passes laws and then the President signs them into law.
- Once an environmental law is passed, an agency (normally the US Environmental Protection Agency (EPA) develops regulations to enforce provisions of the law.
- The agency follows provisions of the Administrative Procedures Act to develop the regulations.

Administrative Procedures Act (1946):

- This act allows the agency to develop regulations that have the force of law.
- The agency will collect information and go through the process of issuing proposed regulations published in the Federal Register.
- Once the regulations are published, time is allowed for public comment. The general public (citizens), public interest groups, industries and anyone else can provide comments to the agency on the regulations.
- Final regulations can be challenged in Federal Court.

What happens after the public comment period?

- The agency must accept all comments and evaluate the merits of the comments in light of the proposed regulations.
- The agency has the option to change the regulations based on public comments, issue them as proposed or withdraw them.

Selected Federal Laws:

- *1. Toxic Substances Control Act (TSCA) —1976
  - Requires the reporting of adverse affects of chemicals.
  - Whenever a company identifies some effect of a chemical's toxicity, impact on the environment or similar information, that was **not previously known**, they are required to report the information to the EPA.
- *2. Resource Conservation and Recovery Act (RCRA) – 1978
  - Defines hazardous wastes, treatment and disposal requirements.
- *3. Comprehensive Environmental Response, Compensation & Liability Act (CERCLA) —1980
  - Established Superfund process. Intended to address existing environmental sites and remediation requirements.

Selected Federal Laws, continued

- *4. Coastal Zone Management Act—1972
  - Sets up process for protection of coastal areas - Texas does not have a separate program. The protection of Texas coastal areas is addressed by other state and federal programs.
- *5. Emergency Planning & Community Right-To-Know Act—1986
  - Provides for identification and communication of the presence of chemicals in local communities and the process of local governments to establish procedures for addressing local issues.

Selected Federal Laws, Continued

- *6. Oil Pollution Act—1990
  - Established procedures for reporting and responding to oil spills.
- *7. Safe Drinking Water Act-1977
  - Provides for standards for drinking water and also a process for local communities to obtain loans to establish or improve treatment systems.

Now let's looks more closely at a few of the more significant laws:

Rivers and Harbors Act of 1899

- The oldest environmental federal law is The Rivers and Harbors Act of 1899.
- The Act made it a misdemeanor to discharge refuse matter of any kind into the navigable waters of the United States without a permit.
- The Act also requires a permit to excavate, fill, or change the course, condition or capacity of any port, harbor, channel, or dam.
- Today this act is still enforced. Permits are issued by the US Corps of Engineers. The permit application process provides for review of the project by various local, state and federal agencies.

Clean Water Act

- The Clean Water Act provides for the protection of water quality.
- The Act is based on the Federal Water Pollution Control Act of 1948 as amended in 1972, 1977 and 1987.
- The revisions in 1972 provided for creation of the National Pollutant Discharge Elimination System (NPDES) permit program.
- Prior to 1972 only states issued water discharge permits.

NPDES Program, continued

- The NPDES permit program is intended to regulate "point sources of pollution".
  - Industrial discharges (point source discharges)
  - Municipal discharges
  - Animal feedlots
  - Non-Point Discharges (streets, parking lots, etc.)

Technology-Based Standards

- The NPDES permit program is a federal program; however, the enforcement and permitting has been delegated to many states including Texas.
- As an example, in Texas the program is administrated by the Texas Commission on Environmental Quality.
- Whenever an application is filed with the agency, the application will include water quality data for the existing or proposed discharges.

Required Data

- The data will include:
  - Detailed chemical analyses of discharges of a wide range of components. Many of the analyses are in the parts per billion (ppb) range.
  - Flow rates.
  - Temperature.
  - Total dissolved solids
  - pH
  - Biological oxygen demand
  - Chemical oxygen demand
  - Metals
  - Inorganic components
  - Organic components
  - Discharge locations
  - Description of industrial process
  - Treatment technology

Permitting Process

- Once the application is accepted by the permitting agency, other agencies are provided the opportunity for comment. The general public will also be allowed to comment.
- The permitting process will normally take a year or more to complete.
- The permit preparation activities (collecting data and preparing the application) can take two or more years.
- Permits normally are issued for a 5 year period and then have to be renewed. The whole process is repeated.

Permit Compliance:

- Once a permit is issued, the permit holder must demonstrate compliance.
- Compliance consists of sampling and analyzing discharges.
- Submittal of monthly discharge reports (DMR's).
- Reporting of any events that do endanger or potentially endanger the quality of the receiving waters.
- Allow inspections by agency personnel.
- There are civil and potential criminal impacts for failure to achieve compliance.

How are the water quality standards established?

- Permit limits are normally based on technology standards, that is, what treatment technology is available?
- If the specific receiving water is impaired, additional requirements can be imposed on the discharge permit.
- The quality of the receiving waters is **not to be degraded**.
- Some regulations are based on the type of industry involved.

Clean Air Act:

- The first federal legislation for **air quality** was the 1955 Air Pollution Control Act. Additional laws were passed in 1963, 1967, 1970, 1977, and 1990.
- The 1990 air pollution control act was the first federal law that provided for permitting of air emission sources.

Clean Air Act, continued

- Prior to changes in 1990, states provided for air permits. Many states now have the authority to issue the required federal permits.
- In Texas, facilities built before a certain date were allowed to continue to operate under "grandfather status" as per state law, but the requirements under the current federal law require permits for all sources.

Clean Air Act, continued

- In addition to air permits, facilities are normally required to submit detailed air emissions records for all of their sources. This inventory includes amounts of pollutants (pounds), volumes, temperatures and identification of pollutants.
- Monitoring of the potential impact of air quality on and off the plant site is a normal activity.
- Impact can be related to human health, wildlife, plant life and facilities.

Clean Air Act, continued

- Permit limits are typically based on Best Achieved Control Technology (BACT). The BACT standards are based on the performance of commercially available control technology. This is a moving target, so when a new facility goes through the permitting process, the proposed controls are compared to the latest technology available for that specific industrial process.
- In some cases Maximum Achievable Control Technology (MACT) is used.
- For example, when a new gas fired power plant is proposed, the agency will review the application and compare the control technology to the **most recent similar facility permitted**.

Clean Air Act, continued

- Air dispersion modeling is typically required.
- Modeling includes list of pollutants.
- Concentrations.
- Location of sources.
- Stack height.
- Flow rates.
- Temperature of releases.
- Local climatic information
  - Wind patterns
  - Description of land features
  - Weather patterns, etc

Clean Air Act, continued

- The regulations and permits require regular reporting of air emissions including any upset events that potentially exceed the requirements.
- Detailed record keeping is required and subject to inspection by agency personnel.
- Allow inspection by agency personnel.
- Civil and criminal requirements can be enforced.

Environmental Management:

- How in the world does a company keep up with all the legal and regulatory requirements?
- It is not easy!
- A well developed **management system** is necessary.
- A system that tracks current and pending requirements is necessary.
- A **well trained staff** and other resources are necessary to achieve and maintain compliance.

- The management system must be **proactive**, that is, a detailed plan that includes an identification of regulatory and permitting requirements:
  - Reporting requirements
  - Due dates
  - Advance timing needed or application preparations
  - Record keeping
  - Monitoring requirements
  - Training requirements
  - Audit process

What is an environmental professional expected to know?

- Basic environmental laws.
- Basic environmental regulations.
- Sampling requirements for air, water, solid wastes, soil and groundwater.
- Analytical methods necessary for determination of potential pollutants.
- Communication skills (written and oral).
- Report preparation.

What is an environmental professional expected to know? (continued)

- Control technology for industrial processes.
- Auditing practices.
- Engineering technology related to processes and controls.
- Chemistry.
- Biology.
- Ecology.

Do Not Be Intimidated If You Are Interested in a Career in Environmental:

- Like any other profession, you prepare yourself with the right education.
- Learn on the job.
- Develop a network of resources.
- Work hard.

EXERCISE 32

A Soldier's Perspective on Environmental Stewardship

Open the aperture – see the big picture.

– Colonel Rick Hoefert (Retired), Former Director, Army Environmental Programs

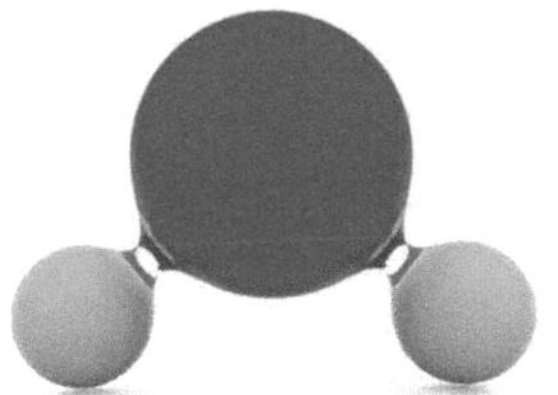

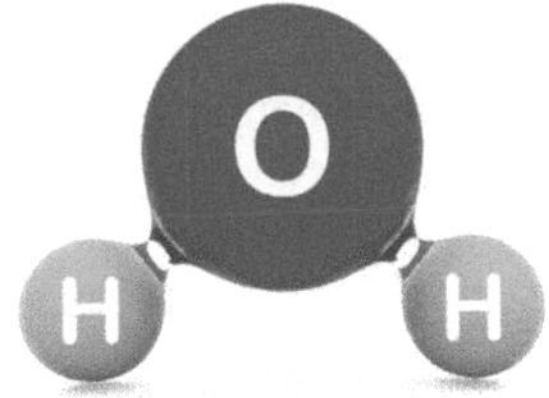

Figure 32.1 The water molecule.

INTRODUCTION

Based on previous exercises in this text, we could conclude that the science of proper care of the environment appears very straightforward.

- Don't pollute.
- Do recycle.
- Don't waste resources.
- Do reuse items.
- Etc.

But is there more to successful implementation of environmentally sound practices? The situation can be complex. Even the simple water molecule, shown in Figure 32.1, can be harmful under certain conditions. Even when all parties have a sound knowledge of the science involved, sometimes conflicts of interest arise, with gray areas where decisions are difficult. A situation like this arose when

DOI: 10.1201/9781003006565-36

the U.S. Army had planned training in areas inhabited by two endangered species, specifically the golden-cheeked warbler and the red cockaded woodpecker. Concerned environmentalists pointed out that we can't have munitions being fired - even if it is just for training - and expect these species to remain safely in the area. Clearly, there was a need for another element - diplomacy.

Should this training, which was designed to bolster the U.S. defense, be halted? Or should the training proceed, with risk to endangered species? What would *we* do if we were tasked with finding a solution to this problem?

In this lab exercise, Colonel Rick Hoefert, Retired, U.S. Army, details how the matter was handled. The Army has emphasized its care for natural resources over the years, as Figure 32.2 shows. And yes, because some people cared about the environment and voiced their concern, and because the military listened, *two birds* influenced the training practices of the world's most powerful military!

INSTRUCTIONS

Read the PowerPoint presentation and answer the questions. Your instructor may discuss this material in class.

EXERCISE 32 A SOLDIER'S PERSPECTIVE ON ENVIRONMENTAL STEWARDSHIP

Last Name____________________ **First Name** ____________________
Instructor ____________________ **Date** ____________________

Instructions: Answer the following based on the information provided in the PowerPoint presentation.

1. State the three main takeaways in managing the environment (or in any professional endeavor).

2. State the practical mandate for sound environmental stewardship.

3. State your answer to Question 2 in one word.

4. Name the two pressures on the US Army's sustainability that are specifically addressed in the presentation.
   a. ____
   b. ____

5. Urbanization/encroachment around Army installations has, in many cases, created:
   a. ____
   b. ____

6. What special day does the US Army promote which supports the sustainability mandate?

7. A sustainable installation systematically decreases its dependence on:
   a. ______________________________ and
   b. ______________________________

8. The cornerstone for the Army's strategy to achieve sustainability and provide for energy security is the Army ______________________ Initiative.

9. As part of this initiative, West Fort Hood, TX, has a ______________ ______________ farm.

10. The facility in question 10 is coupled with ______________________ ______________________ in Floyd County, TX.

11. The features in Questions 10 and 11 together provide ________ % of Fort Hood's electricity.

12. Name the species which was listed as endangered in 1970, and was protected, originally, by restrictions that prevented a soldier from training within 200 feet of a tree cavity which it created.

13. What change, if any, in Army training restrictions in the vicinity of the above species was implemented after wildlife specialists and military planners collaborated?

14. Name the species listed as endangered in 1990, which caused restrictions which forced soldiers to train for combat with "significant artificial workarounds" at Fort Hood, TX.

15. Who were the two lead scientists in the 2009 study to estimate the species distribution and abundance across the species breeding range in Texas for the species referenced in Question 14?
    a. ______________________________
    b. ______________________________

16. List three examples of problems which can occur when passion gives way to emotion.
    a. ______________________________
    b. ______________________________
    c. ______________________________

17. Name the "mystery substance" which is cited as an example of the misuse of science to fear-monger.

18. Name the initiative the new administration of President George W. Bush proposed to reevaluate a number of environmental laws and how they were applied to the military.

19. Name three of the environmental laws proposed to be reevaluated in the initiative referenced in Question 18.
    a. ______________________________
    b. ______________________________
    c. ______________________________

20. How many of the environmental activists attending the Pentagon luncheon who complained about the initiative mentioned in Question 18 had actually read the proposed plan?

Extra Credit (Possible Discussion Question). How do you recommend achieving the proper balance between environmental protection and military readiness?

Figure 32.2 Army Earth Day 1998 poster (www.aec.army.mil).

A SOLDIER'S PERSPECTIVE ON ENVIRONMENTAL STEWARDSHIP

RICHARD A. (RICK) HOEFERT, MSE, MSS
Colonel, Engineer
United States Army (Retired)

DISCLAIMER: The views expressed in this presentation are the author's alone and should not be construed as representing an official position of the United States Army or the United States Department of Defense.

ABOUT THE AUTHOR

Richard A. Hoefert
Colonel, Engineer
United States Army Retired

Served the United States Army for over four decades: 30 years in uniform and another 10 plus years as an Army Civilian

Commissioned as a second lieutenant of Engineers from the United States Military Academy in 1975 with a Bachelor of Science (Civil Engineering concentration)

Holds a Master of Science in Engineering (Civil Engineering Structures) from the University of Texas at Austin

Holds a Master of Strategic Studies from the United States Army War College

Former Director, Army Environment Programs
Headquarters, Department of the Army

Former Chief of Training Support
Headquarters, III Corps and Fort Hood

B.L.U.F.

Bottom Line Up Front

The three main takeaways in managing the environment:
[or in any professional endeavor]

1. **Open the aperture — see the big picture**

2. **Let the science rule the day — not personal agendas**

3. **Be passionate — but not emotional**

B.L.U.F.

Bottom Line Up Front

The three main takeaways in managing the environment:
[or in any professional endeavor]

1. **Open the aperture — see the big picture**
2. **Let the science rule the day — not personal agendas**
3. **Be passionate — but not emotional**

SEEING THE REALLY BIG PICTURE

The altruistic mandate for sound environmental stewardship:

Stewardship = taking care of what belongs to someone else

SEEING THE BIG PICTURE

The practical mandate for sound environmental stewardship:

Use it up now = Nothing left for the future

In other words

We must ensure present requirements are met without compromising the ability of future generations to meet their own needs

In one word

SUSTAINABILITY

PRESSURES ON ARMY SUSTAINABILITY

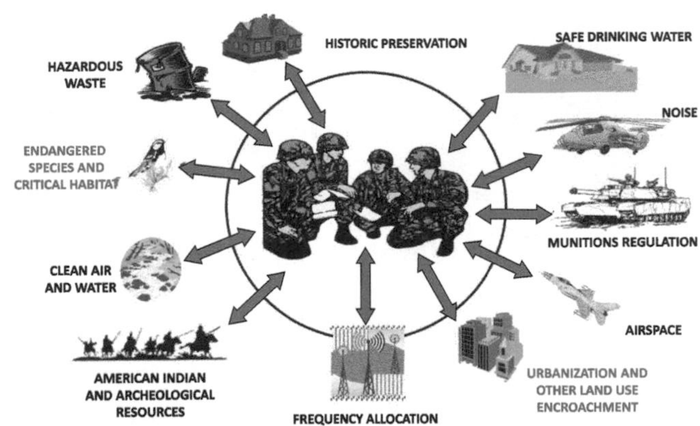

URBANIZATION/ENCROACHMENT (FORT LEWIS, WA)

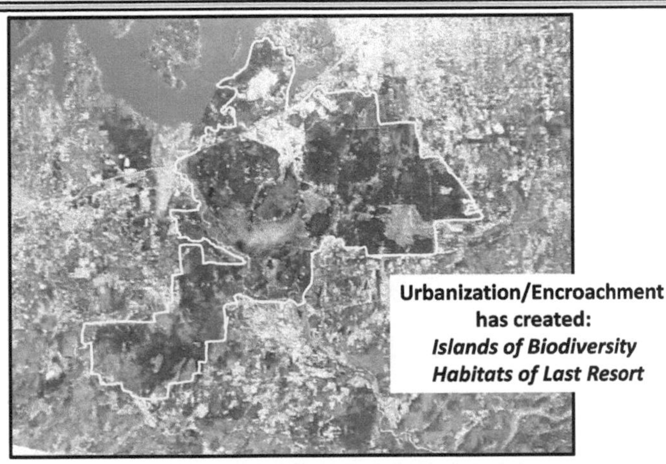

EVOLUTION OF DEALING WITH THE PRESSURES

[NET ZERO]

☺☺ Sustainable Operations/Installations

☺☺ Environmental Management Systems

☺ Pollution Prevention: Pursuit of Excellence

😐 Acceptance = Compliance

☹ Tolerance

☹☹ Denial

THE KEY — SUSTAINABLE OPERATIONS/INSTALLATIONS

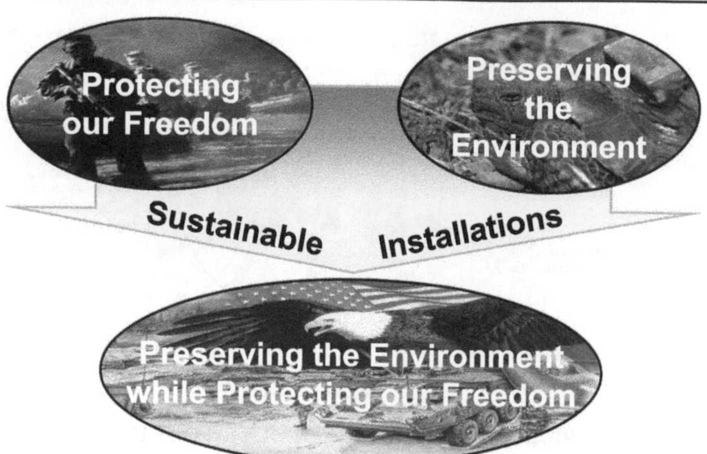

PROGRAMMING (THE $$$) SUSTAINABLE OPERATIONS/INSTALLATIONS

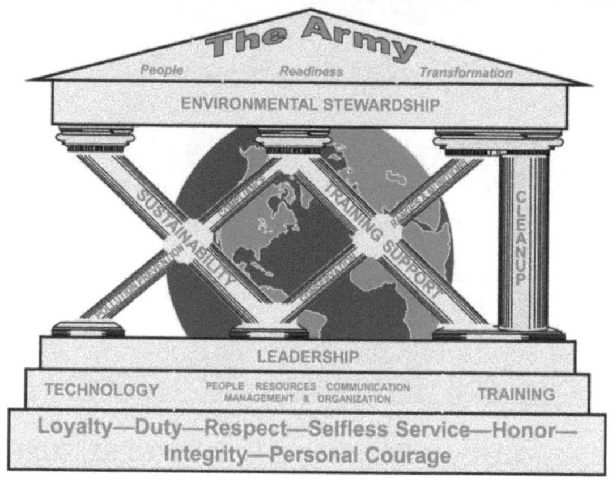

PERPETUATING THE SUSTAINABILITY MANDATE

A SUSTAINABLE INSTALLATION...

- **Fully enables military training**
- **Protects the well-being of Soldiers and Families**
- **Has a mutually-beneficial relationship with the local community**
- **Is life-cycle cost-effective to operate**
- **Systematically decreases its dependence:**
  - *on mined and fossil fuels*
  - *on non-biodegradable and toxic compounds*
- **Does not use resources faster than nature can regenerate them**
- **Operates within its "fair share" of the Earth's resources**

NET ZERO

ARMY NET ZERO INITITATIVE

ARMY NET ZERO INITITATIVE

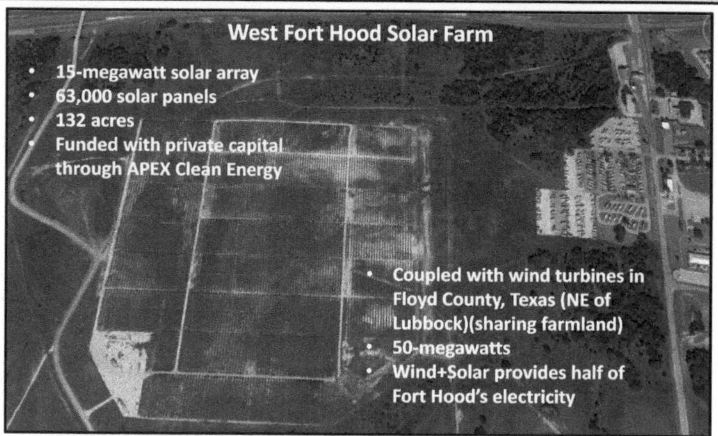

Army overall has 300-megawatts of solar production toward goal of 1-gigwatt by 2025

B.L.U.F.

Bottom Line Up Front

The three main takeaways in managing the environment:
[or in any professional endeavor]

1. Open the aperture — see the big picture
2. **Let the science rule the day — not personal agendas**
3. Be passionate — but not emotional

LET THE SCIENCE RULE THE DAY

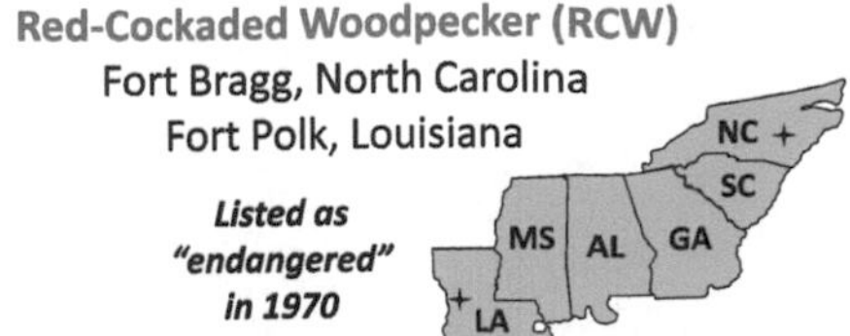

Red-Cockaded Woodpecker (RCW)
Fort Bragg, North Carolina
Fort Polk, Louisiana

Listed as "endangered" in 1970

Golden-Cheeked Warbler (GCW)
[a.k.a. Gold Finch of Texas]

Fort Hood, Texas

Listed as "endangered" in 1990

FORT BRAGG TRAINING RESTRICTIONS

Within 200 feet of a red cockaded woodpecker cavity tree, a Soldier:

- could only walk through the area
- could not dig
- could not use a generator
- could not be in the middle of a fire fight with blanks
- could not use pyrotechnics or smoke
- physically had to stop training

DEMONSTRATING THE CHALLENGE

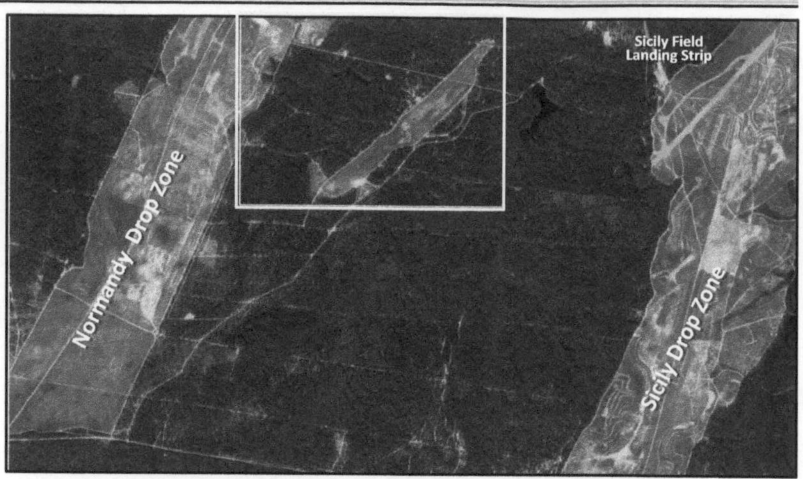

DEMONSTRATING THE CHALLENGE

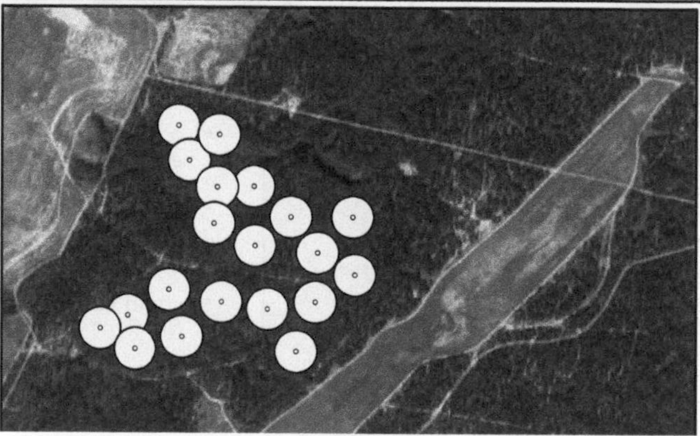

Notional cavity trees with 200-foot restrictive area shown to demonstrate the challenge for Soldiers trying to train in the area between the drop zone and the open area

SCIENCE + COOPERATION = RECOVERY

Training restrictions were relaxed as wildlife specialists and military planners began to understand how Soldiers and equipment in the woods could co-exist with the birds.

One Fort Bragg official said,

"We're no longer adversaries. It's really just about managing the training lands properly."

THE CASE OF THE GOLDEN-CHEEKED WARBLER

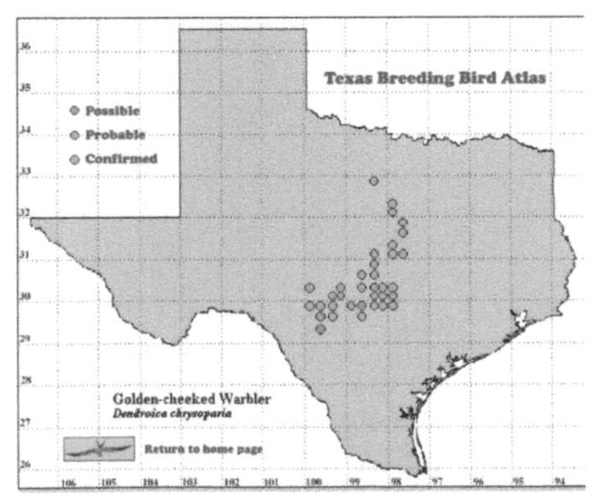

FORT HOOD TRAINING RESTRICTIONS (2003)

- Digging, tree or brush cutting, and "habitat destruction" prohibited throughout the year on 33% of the training areas

During breeding season (March – August):

- Vehicle and dismounted maneuver restricted to established trails
- Site occupations limited to two hours
- Artillery firing & smoke generation prohibited w/in 100 meters of the boundaries of the designated "core areas" (46,620 acres)
- Use of camouflage netting and bivouac were prohibited
- Restrictions forced Soldiers to train for combat with "significant artificial workarounds"

FORT HOOD TRAINING RESTRICTIONS (2005)

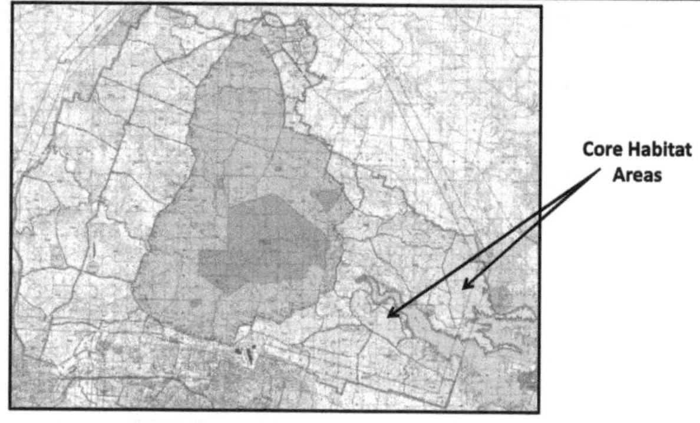

- Restrictions lifted in all but the newly contracted core habitat areas
- Improved species management cited for lifting of the restrictions elsewhere
- Biological Opinion held Fort Hood responsible for >70% of the known Golden-Cheeked Warbler population

SCIENCE TO THE RESCUE???

- In 2009, Texas A&M's Texas AgriLife Research and the Institute of Renewable Natural Resources (IRNR) contracted by Texas Department of Transportation to estimate golden-cheeked warbler distribution and abundance across the species breeding range in Texas.
- Project led by Dr Michael Morrison and Dr Neal Wilkins
- Study results underwent a peer review to determine whether the results are defensible for use as "best available science"
  - "The overall study design, analyses, and inferences are supported by sound scientific data and analysis."
  - "Data collection and statistical procedures were appropriate given the scale and scope of the project."
  - "The methods used for developing and validating the various models are scientifically sound."
  - "Consistently, reviewers felt that this could be used as 'the best available science' for this species."

SCIENCE TO THE RESCUE???

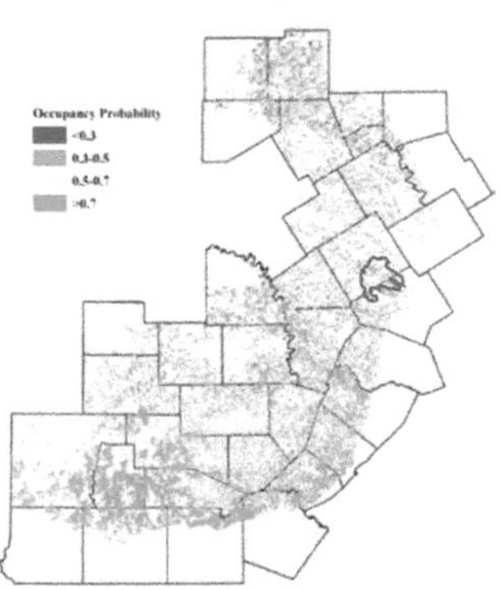

- Just prior to listing in 1990, GCW population estimated at 9,600 to 32,000 birds across its range
- TAMU/IRNR estimated that there were between 175,000 and 265,000 (mean = 220,000) adult male warblers during the breeding season in Texas during 2009
- TAMU/IRNR estimated that Fort Hood had <5% of the GCW population

Despite the new "best available science" little was changed in the management practices for the golden-cheeked warbler

RCW VS GCW—MY CONCLUSIONS

- Every coin has two sides. In this case they are:
- Obverse: The Army and its mission to train Soldiers to fight and win our Nation's wars
- Reverse: Departments, Services, Agencies, Groups, and Societies and their common mission to protect and enhance the biodiversity of Planet Earth
- RCW: Two sides found a common objective to protect the species while allowing Soldiers to train for war
- GCW: Two sides had/have diametrically opposed objectives:
  - Obverse: (adding to the Army others with a vested interest who have joined the fight) Remove training restrictions by de-listing the GCW
  - Reverse: Protect the habitat of the GCW by maintaining the status of the bird as endangered IAW the Endangered Species Act

∴ The saga of the golden-cheeked warbler continues

B.L.U.F.

Bottom Line Up Front

The three main takeaways in managing the environment:
[or in any professional endeavor]

1. Open the aperture — see the big picture
2. Let the science rule the day — not personal agendas
3. Be passionate — but not emotional

BE PASSIONATE—BUT NOT EMOTIONAL

When passion gives way to emotion...

[just three examples to illustrate the point]

1. Science is used to fear monger
2. Decisions are made based on sound bites
3. Judgments are made before getting the facts

SCIENCE IS USED TO FEAR MONGER

Why is this substance not yet banned??

- Substance is colorless, odorless, tasteless
- Kills uncounted thousands of people every year—most caused by accidental inhalation
- Prolonged exposure to its solid form causes severe tissue damage
- Symptoms of ingestion can include excessive sweating and urination, and possibly a bloated feeling, nausea, vomiting and body electrolyte imbalance
- For those who have become dependent, withdrawal means certain death
- Contributes to the "greenhouse effect"
- Contributes to the erosion of the natural landscape
- A major component of acid rain
- Accelerates corrosion and rusting of many metals
- May cause electrical failures and decreased effectiveness of automobile brakes
- Has been found in excised tumors of terminal cancer patients

Dihydrogen Monoxide (DHMO)

DIHYDROGEN MONOXIDE (DHMO)

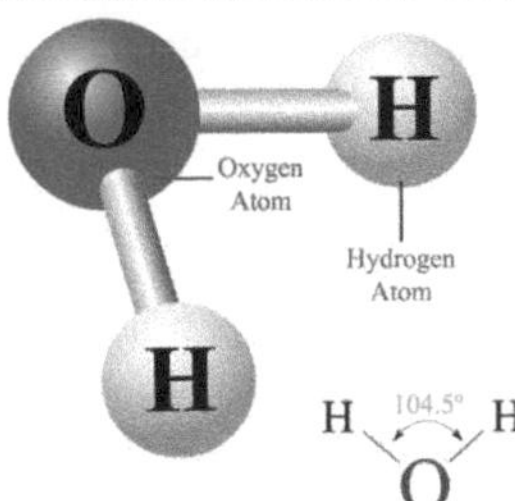

H 104.5° H
O

Dihydrogen = 2 x Hydrogen = 2 H

Monoxide = 1 x Oxygen = O

Or more traditionally

H_2O

DECISIONS ARE MADE BASED ON SOUND BITES

Basic Allowance for Subsistence (BAS)

Sound Bite: "If you stop my husband's separate rations you will be taking food out of my baby's mouth"

Decision: Soldiers' separate rations were no longer stopped upon deployment

The Actual Ledger

- $200	Stopping BAS for the deployed Soldier
+ $100	Cost of food the Soldier would have eaten at home
+ $75	Family Separation Allowance
+ $150	Hazardous Duty/Imminent Danger Pay (a.k.a. Combat Pay)
+ $500	Combat exclusion—pay is not taxed while in a combat zone
+ $625	Change in income during the deployment

** Approximation of the numbers when the decision was made following Operation DESERT STORM*

JUDGMENTS ARE MADE BEFORE GETTING THE FACTS

Readiness and Range Preservation Initiative

Early in the George W. Bush presidency, the Administration proposed reevaluating a number of environmental laws and how they were applied to the Military including:

- **Migratory Bird Treaty Act (MBTA)**
- **Marine Mammal Protection Act (MMPA)**
- **Endangered Species Act (ESA)**
- **National Environmental Policy Act (NEPA)**
- **Clean Air Act (CAA)**
- **Resource Conservation and Recovery Act (RCRA)**
- **Comprehensive Environmental Response, Compensation, and Liability Act (CERCLA)(a.k.a. Superfund)**

JUDGMENTS ARE MADE BEFORE GETTING THE FACTS

Readiness and Range Preservation Initiative

DoD hosted a luncheon in the Pentagon to "socialize" the initiative

Representatives from the Military Departments attended

Representatives from most of the mainstream environmental groups attended

- **Reaction to the initiative was universally negative—condemning the Administration for rolling back the environmental laws**

But NOT ONE of the environmental activists present had actually read the proposed initiative

SUMMARY

The three main takeaways in managing the environment:
[or in any professional endeavor]

1. **Open the aperture — see the big picture**
2. **Let the science rule the day — not personal agendas**
3. **Be passionate — but not emotional**

DISCLAIMER: The views expressed in this presentation are the author's alone and should not be construed as representing an official position of the United States Army or the United States Department of Defense.

END NOTES AND REFERENCES

Slide #1	Although the author is a retired senior Soldier and many of the examples used in the presentation are drawn from military organizations and military installations, the views expressed and conclusions drawn are the author's alone. These views and conclusions should not be construed to represent an official position of the United States Army or the United States Department of Defense.
Slides #3, 4, 16, 28, 29, 35	The three main takeaways were developed by the author for the intended audience of college students undertaking environmental studies. The author derived the main points from four decades of service to the US Army with the main emphasis on two positions held: former Director, Army Environmental Programs and former Chief of Training Support at Fort Hood, Texas.
Slide #6	The definition of "sustainability" was developed by the World Commission on Environment and Development (the Brundtland Commission) in 1987.
Slides #7, 9, 10, 11	Drawn from a presentation by the author at the 2004 Army Environmental Training Symposium. At the time the author served as the Director, Army Environmental Programs, Headquarters, Department of the Army in Washington, DC.
Slide #8	Drawn from a briefing entitled, "Managing for Sustainability: ISO 14001 EMS and Beyond" presented by Paul Steucke, Chief, Environmental Resources Division, Directorate of Public Works, Fort Lewis, WA. August 22, 2002.
Slide #12	Army Earth Day posters are developed by the Army Environmental Command each year. The two posters depicted, as well as the posters for other past years, are available at www.aec.army.mil.
Slide #13	Drawn from a presentation by the author at the 2003 Directorate of Public Works Training Workshop. At the time the author served as the Director, Army Environmental Programs, Headquarters, Department of the Army in Washington, DC.
Slide #14	Graphic drawn from the Army's Assistant Secretary of the Army for Installations and Environment website: https://www.asaie.army,mil/public/es/netzero/.
Slide #15	The image of the Fort Hood Solar Farm was drawn from Google Earth. The data presented are taken from an article in the *Fort Hood Sentinel* newspaper entitled, "Solar, wind provide renewable, secure energy to Fort Hood," by Heather Ashley. June 9, 2017.
Slide #17	https://www.allaboutbirds.org/guide/Red-cockaded_Woodpecker/overview https://www.allaboutbirds.org/guide/Golden-cheeked_Warbler/overview
Slide #18	Training restrictions summarized from United States Corps of Engineers Construction Engineering Research Laboratory Technical Report 99/51 by Larry D. Pater, David K. Delaney, Timothy J. Hayden, Bernard Lohr, and Robert Doding, June 1999, page 12.

END NOTES AND REFERENCES (Continued)

Slides #19, 20	Graphic developed by the author using Google Maps and notional locations for red-cockaded woodpecker cavity trees within the training area of interest to demonstrate the challenges the 200-foot perimeter around each cavity tree presented to Soldiers trying to train in the area.
Slide #21	Data points displayed were drawn from The Fayetteville Observer newspaper article entitled, "Fort Bragg and red-cockaded woodpecker co-exist comfortably after shift in conservation attitudes," by Drew Brooks, August 19, 2014.
Slide #22	Breeding/winter range map drawn from the Wikipedia entry, "Golden-cheeked warbler." Breeding habitat/range map drawn from an Ecosphere article entitled "Spatiotemporal Variation in Range-wide Golden-cheeked Warbler Breeding Habitat," by A. Duarte, J.L.R. Jensen, J.S. Hatfield, and F.W. Weckerly, 2013 (Ecosphere 4(12):152 http://ex.doi.org/10/1890/ES13-00229.1) as attributed to Diamond, et al 2010.
Slide #23	Training restrictions summarized from the U.S. Fish and Wildlife Service's Biological Opinion for Fort Hood, Texas for the federally listed black-capped vireo (*Vireo atricapilla*) (BCVI) and golden-cheeked warbler (*Dendroica chrysoparia*) (GCWA), March 16, 2005.
Slide #24	Training restrictions summarized from the U.S. Fish and Wildlife Service's 2005 Biological Opinion for Fort Hood, Texas. Map is an extract from the Fort Hood Military Installation Map V782S Edition 12. The training staff at Fort Hood used the figure ">70% of the known GCW population" as a summary of the U.S. Fish and Wildlife Service position since Fort Hood had the largest <u>documented</u> population of the species.
Slide #25	The 2009 TAMU/IRNR project is documented in, "Golden-cheeked Warbler Population Distribution and Abundance", by Michael L. Morrison, R. Neal Wilkins, Bret A. Collier, Julie Groce, Heather Mathewson, Tiffany McFarland, Amy Snelgrove, Todd Snelgrove and Kevin Skow (College Station, Texas, July 30, 2010). The peer review of the "Morrison Report" is summarized in The Wildlife Society's, "Review of the Study on Golden-cheeked Warbler Population Distribution and Abundance and Associated Manuscripts—PEER REVIEW," (Bethesda, Maryland, July 18, 2011).
Slide #26	Estimated GCW population prior to listing drawn from, "Balancing Economy with Ecology: A Report to the Legislature from the Interagency Task Force on Economic Growth and Endangered Species," by Susan Combs, Texas Comptroller of Public Accounts, pg 20, (Austin, Texas, November 29, 2010). The map (pg 49) and the estimated GCW male population (pg 107) drawn from the "Morrison Report." The training staff at Fort Hood used the figure "<5% of the GCW population" based on the estimated GCW male population in the "Morrison Report" to reflect that the GCW was more abundant than the previous documented population suggested. The conclusion is the author's based on the management practices at Fort Hood up until the time he retired from Civil Service in 2018.

END NOTES AND REFERENCES (Continued)

Slide #27 The conclusions presented are the author's alone.

Slides #30, 31 en.Wikipedia.org/wiki/Dihydrogen monoxide_parody
https://www.lockhaven.edu/~dsimanek/dhmo.htm

Slide #32 From the earliest time of the Republic, the basis of pay for Soldiers was monetary compensation (currently referred to as "basic pay") on top of the room and board that was provided. This remains the basis for a Soldier's pay to this day. Where the Army cannot provide the actual room and/or board for a Soldier, a monetary allowance for housing (Basic Allowance for Housing) and/or food (Basic Allowance for Subsistence) is provided.
During Operation DESERT SHIELD/DESERT STORM, the Army followed its standard practice of stopping the separate subsistence allowance for the Soldiers deployed as their meals were being provided by the Army. The loss of this income led to a number of complaints from spouses of deployed Soldiers. The separate housing allowance was not stopped for Soldiers with dependents, however, as this allowance was based on providing housing for the Soldier and the family.
Following Operation DESERT SHIELD/DESERT STORM, the Army revised its policy and no longer stopped the Soldier's BAS when the Soldier deployed.
The author personally experienced this new policy during an assignment to the Sinai in Egypt and again when deployed in support of Operation IRAQI FREEDOM.
The numbers used are an approximation only to demonstrate the full picture and not just a single sound bite.

Slides #33, 34 The author was assigned to Headquarters, Department of the Army when the Readiness and Range Initiative was proposed. Later testimony before the Senate Appropriations Committee (fiscal year 2004 Department of Defense Appropriations) on the Readiness and Range Initiative succinctly states the purpose of the initiative:
"The Readiness and Range Initiative (RRPI) provides clarification to specific statutes; it does not provide "sweeping" exemptions from environmental laws... Recently, courts have been interpreting environmental statutes and existing laws in new ways that are impacting military operations on ranges and in airspace."
Shortly after the proposal was made, DoD hosted a luncheon in the Pentagon to lay out the rationale for the proposal and to "socialize" it with the mainstream environmental community. The author was one of the Army representatives at this luncheon and experienced the universally negative reaction to the initiative. When the question was asked as to how many of the environmental representatives had actually read the proposal, not a single hand was raised.

Biographical Notes

Buchner, Ernst (1850–1924): German biochemist who is credited with invention of the Buchner funnel.

Bunsen, Robert (1811–1899): German chemist who invented the Bunsen burner, which is still used in labs today.

Curie, Marie (1867–1934): Influential Polish physicist who made early and significant discoveries about radioactivity. She won the Nobel Prize for Physics in 1903 and the Nobel Prize for Chemistry in 1911.

Curie, Pierre (1859–1906): Influential French physicist who made early and significant discoveries about radioactivity. He won the Nobel Prize in 1903.

Dalton, John (1766–1844): English scientist credited with the initial formulation of modern atomic theory.

Erlenmeyer, Emil (1864–1921): German chemist who invented the Erlenmeyer flask, commonly used in Environmental Chemistry labs.

Fleming, Alexander (1881–1955): Scottish microbiologist and physician who discovered the world's first broad-spectrum antibiotic penicillin, noticing the mold's activity on bacteria.

Franklin, Benjamin (1706–1790): American scientist and founding father.

Gauss, Carl Friedrich (1777–1855): German mathematician and physicist for whom the gauss unit for magnetic field strength is named.

Haber, Fritz (1868–1934): German chemist who developed a reaction that is still used today, to produce ammonia for fertilizers and explosives, from atmospheric nitrogen.

Kekule, August (1829–1896): German organic chemist who recognized that ring-shaped compounds can form, correctly explaining the structure of benzene.

Lavoisier, Antoine (1743–1794): French scientist who is considered by many to be the father of modern chemistry. He investigated and explained combustion, enabling him to aid the French army in its manufacture of explosives and to assist the American colonies with gun powder for their fight for independence from Britain.

Lavoisier, Marie (1758–1836): French wife of Antoine Lavoisier who assisted him with his lab work and is considered by some to be the mother of modern chemistry.

Planck, Max (1858–1947): German physicist who won the Nobel Prize in 1918. A number of prestigious research institutes named for Max Planck operate in Germany today.

Tesla, Nikola (1856–1943): Serbian-American inventor for whom the tesla, a unit of magnetic field strength, is named.

Thomson, William, aka Lord Kelvin (1824–1907): British mathematician and physicist for whom the Kelvin temperature scale is named. Note that Kelvins are not degrees – they are simply Kelvins.

Glossary

Acid anhydride: Chemical compound which forms an acid on reaction with water, such as sulfur trioxide forming sulfuric acid.

ALCOA: Aluminum Corporation of America.

Alloy: Mixture of elements that have been combined in specific proportions, such as brass which contains copper and zinc or dental amalgams which contain mercury, silver, tin, and copper.

Atom: The smallest amount of an element which can exist.

Balance: Device, electronic in most labs, used to determine mass.

Battery acid: Sulfuric acid (H_2SO_4).

Biomass: Renewable form of energy from biological sources such as algae, corn, etc.

Biodegradable: Can be broken down to simpler susbstances by natural environmental processes such as bacterial action.

Biological Oxygen Demand (BOD): Amount of oxygen required to break down pollutants in natural waters, high values being indicative of poor water quality. Carbon cycle Shows by a sequence of reactions how various chemical forms of carbon are used and reused in the environment.

Carbon footprint: The quantities of carbon dioxide and other greenhouse gases which industries or individuals release to the environment due to their daily activities.

Catenation: Ability of atoms such as carbon to bond with each other, forming long chains.

Cetyl alcohol: Chemical compound which can be added to reservoirs to decrease evaporation by forming a film over the waters.

Closed system: System for which the amount of matter present does not change, although the amount of energy can change, such as a capped water bottle in a sunny yard.

Compound: Two or more elements bonded together in definite proportions, such as magnesium chloride ($MgCl_2$).

Corrosion: Breakdown, crumbling, and loss of structure of a substance, such as many metals in the presence of acid.

Data: Information obtained by experiment.

Decant: To pour the liquid off carefully, while leaving the solid matter in the original container.

Deionized water: Water which has been purified by passing it through an ion-exchange column. This process removes ions but dissolved gases remain. Equivalent to distilled water for most lab purposes.

Density: Ratio of mass divided by volume.

Diffusion: Motion of a chemical from an area of greater concentration to an area of lesser concentration.

Distilled water: Water which has been purified by boiling to convert it to a gas, and then allowing it to condense back to the liquid state. This process removes not only ions but dissolved gases as well. Equivalent to deionized water for most lab purposes.

Ductile: Capable of being drawn into a wire, a property of metals such as copper.

Element: Collection of atoms all having the same atomic number (number of protons), e.g., hydrogen (H) with 1 proton or lead (Pb) with 82 protons.

Electrolytes: Substances which conduct electrical current when melted or dissolved in water.

Endothermic reaction: A process requiring energy, such as the electrolysis of water to produce hydrogen (a fuel) and oxygen.

Energy: The ability to do work.

Equilibrium: Process in which the rate of the forward reaction equals the rate of the reverse reaction, resulting in constant concentrations of both products and reactants.

Ester: Product of a reaction of an alcohol and an organic acid.

Flocculant: A compound such as aluminum hydroxide which causes suspended matter to aggregate for easier removal, e.g., in water treatment plants.

Eutrophication: Overgrowth of algae, etc. in natural waters, leading to deterioration of water quality; may be caused by excessive phosphate concentrations in the water.

Evaporation: Conversion of a liquid which is below its boiling point, to the vapor state.

Exothermic reaction: Process that produces energy, such as burning a log in the fireplace.

Extra low frequency (ELF) radiation: Radiation with very low frequency such as 3 sec^{-1}–30 sec^{-1} and very long wavelength such as 10^4–10^5 km; often produced by power lines or electrical devices.

Extraction: Use of a solvent to separate a substance from a mixture, such as pesticide from beeswax.

Extrinsic property: A property of matter such as mass or volume, which varies with the sample size.

Filtrate: The clear liquid which is separated from solid matter and which flows through the stem of a funnel during filtration.

Galvanized steel: Steel coated with zinc as protection from corrosion.

Gasohol: Vehicle fuel to which ethanol has been added.

Gauss: Magnetic flux density unit.

Greenhouse gases: Molecules such as carbon dioxide, water, CFCs, and methane, which help trap the sun's energy on the earth, contributing to rising temperatures.

Haber reaction: Synthesis of ammonia from hydrogen and atmospheric nitrogen; very useful for the production of fertilizer and explosives.

Heat: Energy transfer between objects at different temperatures.

Heat of reaction: Quantity of energy released in or required for a chemical reaction.

Heat of solution: Quantity of heat produced or absorbed when a substance sdissolves in a solvent.

Heterogeneous: Varied concentration throughout the sample.

Homogeneous: Same concentration throughout the sample.

Hydrolysis: Chemical reaction involving water to break down a molecule to simpler substances, such as acid-catalyzed hydrolysis of proteins to amino acids.

Hydrophilic: Attracted to water, polar.

Hydrophobic: Not attracted to water, non-polar.

Hydrosphere: The total amount of the Earth's water - as oceans, rivers, glaciers, lakes, etc.

Intrinsic property: A property of matter such as density, which is independent of the size of the sample.

Isotopes: Different forms of the same element, having the same number of protons but a different number of neutrons.

Limestone rock: Compressed calcium carbonate ($CaCO_3$).

Malleable: Capable of being pounded into a thin sheet, e.g., gold is very malleable.

Mass: Actual amount of matter present in a sample; does not vary with location.

Matter: Something that takes up space and has weight, e.g., water is matter whether it is in the liquid, gaseous, or solid state.

Metals: Elements with high levels of ductility, luster, malleability, and electrical and heat conductivity. Usually silver or gray in color, copper and gold being notable exceptions. Usually react by the loss of electrons, a process which is termed oxidation.

Mixture: Two or more substances which are combined but not chemically bonded to each other.

Muriatic acid: Hydrochloric acid (HCl).

Monomer: Repeating unit in a polymer, such as ethylene in polyethylene.

Neutralization reaction: Chemical change occurring when acid is added to basic solutions, or base is added to acidic solutions with the purpose of adjusting the pH to a safer level such as 7; involves the transfer of hydrogen ions (H^+).

Non-metals: Elements with low levels of ductility, luster, malleability, and electrical and heat conductivity. Tend to have a variety of colors. Usually react by the gain of electrons, a process which is termed reduction.

Non-polar substances: Substances composed of molecules with homogeneous electrical charge distributions, such as all hydrocarbons, e.g., octanes (C_8H_{18}).

Open system: System for which both the amount of matter and the quantity of energy can change, such as an uncapped water bottle in a sunny yard.

Oxidation-reduction reaction: Chemical change due to the transfer of electrons between reacting substances.

Personal Protective Equipment (PPE): Approved lab coat or lab apron and approved splash goggles, possibly with gloves or other items based on the nature of the experiment performed.

Phenolphthalein: Acid-base indicator which is pink in base and colorless in acid.

Plastic: A polymer which can be softened, shaped, and allowed to harden, preserving the desired shape.

Polar substances: Substances composed of molecules with unequal electrical charge distribution, such as hydrogen fluoride in which a slight positive charge is localized on the hydrogen and a slight negative charge on the fluorine.

Pollution: Excessive concentration of a substance or product in an undesirable location, such as ozone in the air humans breathe. Ozone in the upper atmosphere is valuable for its ability to absorb ultraviolet radiation and provide protection against skin cancer and other health problems.

Polymer: Giant molecule composed of many smaller molecules which are linked by chemical bonds.

Polymer, addition: Polymer formed by reaction of molecules with double bonds combining without losing any atoms.

Polymer, condensation: Polymer formed by reaction of molecules which release water molecules.

Precipitate: The insoluble substance which forms when two soluble reagents are combined.

Precipitation reaction: A chemical change in which a solid, insoluble material is formed from solutions of two soluble substances which have been mixed.

Radioactivity: Release of high-energy radiation due to breakdown of atomic nuclei.

Renewable: Applied to resources, indicates they can be replaced, such as ethanol derived from corn.

Saponification: Preparation of soap by reaction of triglycerides (fat) and lye.

Slurry: Saturated solution with undissolved solute present.

Solute: The substance which is dissolved by a solvent to form a solution, such as sugar dissolving in water.

Solvent: The substance used to dissolve a solute and form a solution, such as water dissolving sodium chloride from rocks.

Spectroscopy: Analytical method involving the interaction of a substance with light of specific wavelength(s) for identification and/or quantification (measuring concentrations).

Standard: A chemical which can be obtained in a highly purified form and used to determine the concentration of other chemicals.

Sublimation: Direct conversion of a substance from the solid to the gaseous state, without passing through the liquid state.

Steel: Iron combined with carbon for greater strength.

Temperature: Measurement determined by average kinetic energy (energy of motion) of a sample's atoms and molecules.

Tesla: Unit of magnetic field strength.

Thermochemistry: Branch of chemistry dealing with the energy involved in a reaction.

Vapor pressure: Pressure exerted by a vapor in equilibrium with its liquid.

Viscous: Having a high viscosity, very resistant to flow, such as molasses.

Weight: Attraction of gravity for a sample, which varies with location.

Index

Page numbers in *italics* indicate figures; page numbers in **bold** indicate tables.

abbreviations, in notebook, 32
AC current, 150
accuracy, of glassware, 39–40; Class A/Class B glassware, 40; Erlenmeyer flask, 40, *42*; graduated cylinder, 39–40, *41*; unclassified glassware, 40
acetic acid, 136, 158, 165
acetone, in glassware cleaning, 45
acetylene, 157
acid rain, 58, 83, 84, 103, 164
acid-base reactions, energy from, 129–34
acids, 164–5
actinide series, 104, *105*
activated charcoal, 205; use of, 207
activities to avoid, *19*
Administrative Procedures Act, 255
air, laboratory analysis, 213–221; determination of dissolved solids in, **219**; determination of particulate solids in, **219**; determination of pH of water exposed to, **220**
air quality, 213, *217*
ALCOA. *See* Aluminum Corporation of America (ALCOA)
alkaline solutions, storing, 42–3
alloys, 104. *See also* metals; amalgams, 104; brass, 106; exercises, 109–11; rose gold, 104; solder, 104
alternating current (AC), 150
alternative energy sources, 122
aluminum, *95*
Aluminum Corporation of America (ALCOA), 250
aluminum hydroxide: preparation and use of, 206–7; in removal of suspended solids, 204–5
amalgams, 104
amino acids, 160
ammonium nitrate, 133
amorphous solid, 36
anthocyanins, 163–5; pH and, *163*, 165; structure of, *164*
antibacterial additives in soaps, 180
antioxidants, 164. *See also* anthocyanins
aquatic plants, detergent concentrations and, 179–80
Army Earth Day, *268*
Atomic Theory, 144

background radiation, 144
balance: centigram, *1*; measuring mass, 90, 91
balancing chemical equations, 82–3
bananas, background radiation, 144, **146**
barium hydroxide, 133–4
barium sulfate, 133
bases, 165
"bathtub ring", 82, 180
batteries, recycling, *62*. *See also* recycling
battery acid, 84
beaker, *1*
beaker tongs, *2*
Beard, James M., 203
beeswax, 172, *172*
benzene, *155*
Beral pipet, *3*
beverages: decaffeinated, 171; electrolyte-enriched, *135*
biodegradable material, 224
biological oxygen demand (BOD), 190
BOD. *See* biological oxygen demand (BOD)
boiling point, 121
boron, 37, *38*
borosilicates, 37
brass, 106
Brazil nuts, background radiation, 144, **146**
breakage, protecting glassware from, 39, *40*
bronze, *113*
brushes, for cleaning glassware, 37
Buchner funnel, *4*. *See also* funnel
Bunsen burner, *4*, 31, *39*, 125
buret, 43, *44*
burner, Bunsen, *4*, 31, *39*, 125
burns, 39
butane, 158

caffeine, extraction, 171, *171*, 172, 173
calcium carbonate, 85
calorimeter, 130, 131
candles, exothermic reactions, *129*
"canned heat", 121–8
carbon cycle, 85
carbon dioxide, 18, 71, 173, 196
Carbon Footprint, 78, *81*
carbon monoxide, 83, 172
carbonates, 85
care and cleaning, glassware, 35, 37
career, 261

catenation, 236
Celsius (°C), 71, 90
centimeters, 90
Central Nervous System (CNS) stimulant, 173
centrifuge, *5*
cetyl alcohol, 122
chemical reactions, environmental, 81–86, 130. *See also* specific chemical reactions; balancing, 82–3
chlorination, of swimming pools, 204
chloroform, 159
chlorophyll, 164
cigarettes, 145
Class A glassware, 40
Class B glassware, 40
Clean Air Act, 258–9
Clean Water Act, 256
cleaning, glassware, 35, 36, 44–5, *45*
closed system, 122
Coastal Zone Management Act, 255
coinage metals, *103*, 106, 108
colloid, 122
coloring agents in glass, 37
Comprehensive Environmental Response, Compensation, and Liability Act (CERCLA), 255
conical funnel, *6*
contaminants, errors due to, 36
conversion chart, 61, **62**
cooking fuel, *121*
copper: bronze, *113*; as coinage metal, *103*, 106, 108, 113; conversion of copper (II) sulfate to copper metal, 115–6; recycling, 113–7, **114**; rose gold, 104, 113; sequence of chemical changes, **114**; uses, 113; zinc-coated, 106, 108
copper (II) hydroxide: conversion to copper (II) oxide, 115; copper (II) nitrate to, 115
copper (II) nitrate: conversion of copper metal to, 114–5; conversion to copper (II) hydroxide, 115
copper (II) sulfate: conversion to copper metal, 115–6; copper (II) oxide to, 115
copper wire, 104, 107
copper (II) oxide: conversion to copper (II) sulfate, 115; copper (II) hydroxide to, 115
cork ring, 39, *40*
corrosive reactions, 107, 108, 116, 125, 131
cosmic radiation, 145, **147**
covalent compounds, 138
COVID-19 pandemic, soap use and, 179
cross-linked polymer, 224, 225, 227
"crosslinking" polymer, 225, 227
Curie, Marie, 144
Curie, Pierre, 144
Cuyahoga River (Cleveland, Ohio), 203

Dalton, John, 144
data: falsifying, 25; table, 31, *31*
data table, sample, 31, *31*
DC current, 150
DDT (dichlorodiphenyltrichloroethane), 156
degrees Celsius (°C), *71*, 90
degrees Fahrenheit (°F), *71*, 90
dehydration reaction, 115
deionized (DI) water, 44, 116
density: classification of plastics by, **238**; column, 96, *96*; defined, 56, 96; exercises, 97, 99–101; formula, 96; high-density polyethylene, 95; as intrinsic property, 96; lead, *235*, 236; of a liquid, 98, **99**; low-density polyethylene, 95; mercury, 95; of a solid, 98, **99**
depression, 149
detergents: for cleaning glassware, 37, 45; defined, 180; environmentally friendly, *189*; micelle formation, *181*; phosphate ions in, 189–94; synthetic, 180
diffusion, 249
direct current (DC), 150
disposal, of Styrofoam™, 224
dissolved chemicals, 205
dissolved solids: in air, **219**; in water, 196, **199**
distilled water, 44
DMF. *See* N,N-dimethylformamide (DMF)
donating and receiving electrons, 138
drying, glassware, 45
ductile metals, 104

electricity, ions and, 136
electrolysis, 204
electrolyte-enriched beverages, *135*
electrolytes, 90, 135–42; concentrations and marine life, 135–36; conductivity device with, *137*; exercises, 139–42; strong, 136; weak, 136
electromagnetic frequency (EMF), 150
electromagnetic spectrum (EMS), *149*, 149–50
electronic notebooks, 32, *32*. *See also* laboratory notebook
electrons: donating and receiving, 138; sharing, 138
ELF. *See* extremely low frequency (ELF) electromagnetic radiation
Emergency Planning and Community Right-to-Know Act, 255

EMF. *See* electromagnetic frequency (EMF)
EMS. *See* electromagnetic spectrum (EMS)
endothermic reactions, 130
energy: from acid-base reactions, 129–34; defined, 130; enthalpy change (ΔH), 130–1; first law of thermodynamics, 130–1
enthalpy change (ΔH), 130–1. *See also* energy
Enviro-Bond™, 224, 225, 227
environmental crime, 250
environmental law, 249–61
environmental legislation, 249
environmental management, 250, 260
environmental pollutants, 61–62. *See also* parts per million (ppm)
environmental stewardship, soldier's perspective, 263–85
EPA. *See* US Environmental Protection Agency (EPA)
equipment, safety, 18–19. *See also* safety, environmental chemistry lab
Erlenmeyer, Emil, 40
Erlenmeyer flask, *8*, *39*, *44*; accuracy, 40, *42*
errors: contaminants, due to, 36; glassware, determining, 40–1; parallax, 98
esters, 159
ethane, 157
ethanol, 122, 125, 158, 215
ethics, environmental chemistry lab: plagiarism, 26; post-lab questions, 26–7; research and science ethics crossword, *25*; sectors of community life and, *26*
ethyl acetate, 159
ethyl alcohol, 122, 125
ethylene, 157, 223
eutrophication, 179–80, 189–90
evaporation, 121–2
exercises, in environmental chemistry lab: air, laboratory analysis, 218–21; chemical reactions–balancing equations, 84–6; density, 97, 99–101; electrolytes, 139–41; ELF radiation, 151–4; energy from acid-base reactions, 134; environmental stewardship, 265–8; evaporation and "canned heat", 124–8; extraction process, 175–7; introduction to glassware, 46–50; law, 251–2; measurement, 92–4; metals and alloys, 109–12; notebook, 33–4; organic chemistry, 157–60; parts per million (ppm) and other tiny quantities, 60–2; pH, 168–70; phosphate measuring with spectroscopy, 192–4; polymers, 228–33; radiation, 146; recycling, 240–4; recycling copper, 117–9; safety, 22–4; soap/detergent, 184–8; water, laboratory analysis, 198–201; water treatment process, 208–11
exothermic reactions: candles, burning, *129*; defined, 130
exposure, forms of radiation, **147**
extraction process, 171–77; beeswax, 172, *172*; caffeine from tea, extraction, 171, *171*, 172, 173; exercises, 175–7
extremely low frequency (ELF) electromagnetic radiation, 149–54, *154*; exercises, 151–4; measurement near power lines, *154*
extrinsic property, 96

Fahrenheit (°F), 71, 90
falsifying data, 25
fats and oils, soap making from, *179*, 180
fertilizers, phosphates in, 189
filtration process, 215
first law of thermodynamics, 130–1. *See also* energy
flasks: various sizes, *43*; volumetric, 42–3, *43*
Fleming, Alexander, Sir, 45
float, 95, 96, 236
flocculant, 204
foods, background radiation, 144, **146**, **147**
fossil fuels, 68, 122
Franklin, Benjamin, 13
fuels, 122
funnel: Buchner, *4*; conical, *6*; separatory, *7*

galvanized steel, 84
"Gamma Knife", 150
gamma rays, 150
gasohol, 122
gasoline, 122
Gauss, Carl Friedrich, 149, 150
gels, 122
glassware and supplies, *1–9*; accuracy of, 39–40, *41*, *42*; breakage, protecting from, 39, *40*; care, 35, 36; cleaning, 35, 36, 44–5, *45*; composition, 37, *38*; contaminants, errors due to, 36; determining possible error, 40–1; heat protection, 38–9, *39*, *40*; importance of, 35; to measure volume in lab, *85*; scratch protection, 37; typical lab, *14*; volumetric, 42, *43*, *44*, *45*
glycine, 160
gold: as coinage metal, 103, 106; rose gold, 104, 113
golden-cheeked warbler, 264
graduated cylinder, *6*, 90, 91, 98; accuracy of, 39–40, *40*

greasy dirt, 180
greenhouse gases, 71, 78, 81
guide, safety, 16–18. *See also* safety, environmental chemistry lab

Haber synthesis, 84
handheld radiation meter, 144
HDPE. *See* high-density polyethylene (HDPE)
heat, 72–76; calculation, 72–74; conductors of, 104
heat of reaction, 130–1
heat of solution, 43
heat protection, glassware, 38–9, *39*, *40*. *See also* glassware and supplies
heat-resistant product, 37
high energy radiation, 150
high-density polyethylene (HDPE), 95. *See also* density
Hoefert, Colonel Rick, 263, 264, 269–85
homogenous, 183
hydrochloric acid, 84; neutralization reaction of, 132; reaction of pennies with, 106, 107; strong acid, 164
hydrofluoric acid (HF), 165
hydrolysis, 180
hydronium ion, 164, 165
hydrophilic (polar) portion, 180
hydrophobic (nonpolar) portion, 180
hydrosphere, 72
hypertonic solution, 136
hypotonic solution, 136

ice, density of, 96
internal combustion engines, 122
International System of Units. *See* metric system
intrinsic property, 96
invisible pollutants, 143–8
ionic compounds, 138
ions, 136
isolated system, 130
isotopes, 144

Jefferson, Thomas, 25
Jello™, 122
jelly, 122

Kekule, August, *155*, 156
key lab safety reminders, *13*, 14

laboratory glassware and supplies, *1–9*
laboratory notebook, 29–33; electronic, 32, *32*; entering results in, *29*; importance of, 29–30; pointers for a, 30–2; spiral-bound, *30*
Lake Tahoe, *203*
lanthanide series, 104, *105*
Lavoisier, Antoine Laurent, 51, 57, 58. *See also* metric system
Lavoisier, Marie, 57, 58
Law of Conservation of Energy, 249
Law of Unintended Consequences, 179
LCD. *See* liquid crystalline display (LCD)
LDPE. *See* low-density polyethylene (LDPE)
lead, 37, *38*; alloy of, 104; density, *235*, 236; mass of, 236
lead nitrate, *82*
legislation, environmental, 249–50; chemical manufacturing plants, *250*
limestone rock, 85, 86
liquid, density of, 98, **99**
liquid crystalline display (LCD), 104
liquid metal, 104
liters, 90
lithium oxide, 138
litmus paper, *7*
low-density polyethylene (LDPE), 95. *See also* density
lye, 180

magnetic flux density, 150
Magnetic Resonance Imaging (MRI) studies, 144
malleable metals, 104
marine life: detergent concentrations and, 179–80; electrolyte concentrations and, 135, 136
mass, 90; determinations, 96; of lead, 236
Matcha Mole, *80*
Maximum Contaminant Level (MCL), 36, 62
Mayfield, John C., 250
MCL. *See* Maximum Contaminant Level (MCL)
measurements: exercises, 92–4; , glassware used to measure volume, *89*; importance in environmental chemistry, 90; metric system, 51–6
memory metal, 104
meniscus, 96
mercury: alloys of, 104; density, 95; in drinking water, 41
metalloid, 37
metals, 103–12; alloys, 104; coinage, *103*, 106; conductors of heat, 104; ductile, 104; exercises, 109–12; malleable, 98; overview, 103–4; periodic table, 104, *105*
metathetical reaction, 115
meter, 90

meter sticks, *51*
methyl amine, 159
methylene chloride, 173, 174
metric system: concentration units and their definitions, 61, **62**; exploring, 51–6; mastering, 57–60; unit conversions, 52–6
micelle, *181*
microgram, 61
milligauss, 150
milliliters, 90
millirem (mrem) per year, 144
"mini-percentages", 61
mold growing, on glassware, 45, *45*
molecular modeling kits, 156
monomers, 224
MRI. *See* Magnetic Resonance Imaging (MRI) studies
muriatic acid, 84

narrow-necked containers, 43
National Oceanic & Atmospheric Administration, 254
National Pollutant Discharge Elimination System (NPDES), 256, 257
neurotoxin, 37, 95
neutralization reactions, 82, 84, 132
neutrons, 144
NiTiNaval Ordinance Laboratory, 104, 107
Nitinol, 104, 107
nitrite ion, 36
N,N-dimethylformamide (DMF), 172
non-biodegradable material, 224
non-electrolytes, 136; conductivity device with, *137*
non-metals, 104, *105*
nonpolar solvents, 172, 173
non-renewable resources, 224
notebook, laboratory. *See* laboratory notebook
nuclear radiation: applications of, *143*, 144; common exposure, **147**; isotopes, 144; percentages of exposure in our environment, *148*
nucleus, 144

octane, 122
Oil Pollution Act, 256
oil spills, 92, 225
open system, 122
organic chemistry/compounds, 155–61; as pollutants, 156
organic compounds, 156
oxidation-reduction reactions, 82, 84, 115

PAN. *See* peroxyacetyl nitrate (PAN)
parallax errors, 98
parts per billion (ppb), 41
parts per million (ppm), 41, *61*, 61–4
parts per thousand (ppt), 62
pearls, 85
penalties, environmental crime, 250
penicillin, 45
periodic table, metals, 104, *105*
peroxyacetyl nitrate (PAN), 156
petroleum, 122
pH, 165; detergent concentrations and, 179; indicator, red cabbage and, *163*; of water exposed to air, determination of, 216, **219**
phase change, 104, 106, 107
phosphate ions in detergents, 189–94; eutrophication and, 189; measuring with spectroscopy, 189–94; phosphate concentrations of unknown solutions, **193**; phosphate standard solutions, preparation of, **190**; preparation and absorbance readings of phosphate solutions, **191**
phosphate solutions: preparation and absorbance readings of, **191**; standard solutions, preparation of, **190**
pipets, 43, *45*; Beral, *3*
plagiarism, 26
Planck, Max, 89
plant: aquatic, detergent concentrations and, 179–80; color, pH and, 165
plastics, 223. *See also* polymers; recycling; classification, by their number and density, 235, **238**; for recycling purposes, classifications of, *237*; separation of, 244
polar solvents, 173
pollutants, 90, 103; air, 214; invisible, 143–81 (*See also* nuclear radiation); organic compounds, 156
pollution, environmental law, 249–61
polonium-210, 145
polyethylene, 224; structure of, *232*
polymerization reaction, 233
polymers: classification, for recycling purposes, *237*; cross-linked, 224, 225, 227; Enviro-Bond™, 224, 225, 227; monomers, 224; sodium polyacrylate, 226–7, 224; starch pellets, 224, 210; structure, *223*, 223–4; types, 225
post-lab questions, ethics and, 26–7
potassium chromate, *82*
potassium perchlorate, 85
precipitation reactions, 82, *82*, 85

protein, *160*
protons, 144
Pyrex, 37, 39

qualitative studies, 36
quantitative studies, 36

radiation, 143–8; cosmic, 145, **147**; exercises, 145–6; nuclear (*See* nuclear radiation); sources and their amounts in environment, **147**; terrestrial, **147**
radioactive isotopes, 144
recording information in notebook, 32
recycling: aluminum, 204; batteries, *62*; copper, 113–9, **114**; plastics (polymers), 235–44, *237*
red cabbage, *163*
red cockaded woodpecker, 264, 275
rem, 144
reminders, key lab safety, *13*, 14
renewable resources, 224
research and science ethics crossword, *25*
Resource Conservation and Recovery Act (RCRA), 255
ring stand, in heating glassware, 39, *39*
rinsing, glassware, 44–5
Rivers and Harbors Act of 1899, 256
rose gold, 104, 113
rubbing alcohol, 122

SAD. *See* seasonal affective disorder (SAD)
Safe Drinking Water Act, 256
safety, environmental chemistry lab, 13–23; equipment, 18–19; guide, 16–18; key lab safety reminders, *13*, 14; Safety Agreement, 14, 15, 20–1; typical lab glassware, *14*
Safety Agreement, 14, 15, 20–1. *See also* safety, environmental chemistry lab
salt (sodium chloride): concentration in ocean, 90, 136; measurement, 90; structure, 36, *36*, *37*
saponification, 180
scientific notation, 65–9
scientific research, ethics and, 26
scratch protection, glassware, 37. *See also* glassware and supplies
seasonal affective disorder (SAD), 149
separatory funnel, *7*
shape memory alloy (SMA), 104, 107; applications, 104
sharing electrons, 138
SI Unit. *See* metric system
silica, 37
silver: as coinage metal, 103, 106
sink, 95, 96, 238, 239
sintered glass center, heating glassware and, 39, *39*
sizes, flasks, *43*
Sketch Method, 156
SMA. *See* shape memory alloy (SMA)
"soap scum", 180
soaps: COVID-19 pandemic and, 179; disadvantage of, 180; making from fats, *179*, 180; preparation of, 182–3; "soap scum", 180
sodium chloride (salt): concentration in ocean, 90, 136; dry, 136; measurement, 90; structure, 36, *36*, *37*
sodium hydroxide, 108, 132, 180
sodium polyacrylate, 224, 226–7
solder, 104
solid: density of, 98, **99**; phase changes, 104
solute, 42
solution, hypertonic/hypotonic, 136
solvent, 42
solvents: nonpolar, 172; polar, 172
specific heat, 72
spectrophotometer, *7*
spectroscopy, 189–94
spiral-bound lab notebook, *30*
standards, 190; phosphate standard solutions, preparation of, **190**
starch pellets, 224, 226
Statue of Liberty, *57*, 58
steel, 84
Sterno Brand Canned Heat™, *121*, 121–22
stirring rods, for cleaning glassware, 37
strong acid, 164, 165
strong electrolytes, 136
Styrofoam™ (polystyrene), 224
sublimation, 173, 174
sulfur dioxide, 83
sulfuric acid: heat of solution, 43; neutralization reactions, 84
surroundings, 130
suspended solids, 196, 197, **199**, 204; removal, aluminum hydroxide in, 204–5
Sweet'N Low packet, atoms in, *65*
swimming pools, chlorination, 204
symbols, for chemical elements, 77–80
synthetic detergents, 180
system, 130; closed, 122; isolated, 130; open, 122

table use in notebook, 31, *31*
TC pipette, 43
TD pipette, 43

TDS. *See* total dissolved solids (TDS)
tea, extracting caffeine, 171, *171*
Teflon™, 224, *217*
temperature, 71–76; comparison of temperature scales, **72**; conversion, 72–4; Earth, maintaining, 131; equations for changing temperature units, 72; measurement, 86, 87; Nitinol and, 104
terrestrial radiation, **147**
tesla, 150
Tesla, Nikola, 150
tetrafluoroethylene, 224
Texas Attorney General, 254
Texas Commission on Environmental Quality, 254
Texas Department of Health, 254
Texas Historical Commission, 254
Texas Land Commission, 254
Texas Parks and Wildlife Department, 254
thermochemistry, 123–8
thermometer, *39*, 91
Thomson, William, Sir, 51
tongs, beaker, *2*
total dissolved solids (TDS), 186, **199**
total suspended solids (TSS), 196, **199**
Toxic Substances Control Act (TSCA), 255
training restrictions, 264, 275, 278
triangular file, *8*
triclocarban (TCC) in soaps, 180
triclosan (TCS) in soaps, 180
TSS. *See* total suspended solids (TSS)
tumors, Gamma Knife and, 150
typical lab glassware, *14*

unclassified glassware, 40
unit conversions, 52–6
United States, map of water hardness, *182*
uranium oxide, in glass staining, 37
US Bureau of Marine Fishers, 254
US Corps of Engineers, 254
US Department of Justice, 254
US Environmental Protection Agency (EPA), 36, 61, 225, 254
US Fish and Wildlife Service, 254
US Geological Survey, 254

vapor pressure, 122
voltmeter, 136
volume, 58; measurement, 90, 91, 96
volumetric flask, 42–3, *43*
volumetric glassware, 42, *43*, *44*, *45*. *See also* glassware and supplies

washing, glassware, 35, 36, 44–5
watch glass, *8*
water: body composition *1962*, 204; concentration of suspended solids in, 187, **199**; density, 220; dissolved solids in, determination, 197, **199**; laboratory analysis, 195–201; mass, 236; molecule, *196*, *263*; quality of life and, *195*; treatment process, 203–11
water, density of, 95, 96
water hardness in the United States, *182*
water treatment process, 203–11; activated charcoal, 205–6; aluminum hydroxide in, 204–5; water treatment plant, *205*
weak acid, 165
weak electrolytes, 136
weighing boat, *9*
weight, 90
wire gauze, in heating glassware, 38–9, *39*

X-rays, 144

zinc chloride, 116
zinc sulfate, 116
zinc-coated copper, 106, 108

Milton Keynes UK
Ingram Content Group UK Ltd.
UKHW051014071024
449327UK00012B/239

9 780367 438951